Gvozdev:

The Russian Discovery of Alaska in 1732

By: L.A. Goldenberg

Edited by: J.L. Smith, MA

White Stone Press
Anchorage, Alaska
2001

Copyright 2001, White Stone Press
International Standard Book Number: 0962672734
Revised Edition
Library of Congress Control Number: 2001-131407
English translation 2001 by White Stone Press
2314 Marian Bay Circle, Anchorage, AK 99515
All Rights Reserved. Published 2001
Printed in the United States of America

In loving memory of my son,
James L. Smith III

Acknowledgements

As a student of early Alaska history I first became aware of the essence of the story of the Russian discovery of Alaska while taking a correspondence course in Alaska History from the University of Alaska at Fairbanks about fifteen years ago. The more research I did on the subject the more puzzled I became as to why so little accurate information about the discovery was in print in the English language. When I discovered Goldenberg's biography of M.S. Gvozdev in the library of the University of Hawaii my immediate thought was that it should be translated into English and published. The result was the 1990 paperback edition of this book.

My education in early Alaska history continued with the publications of *To the American Coast* in 1997 and *Russians in the Bering Strait, 1648-1791* in 2000. I believe the time has come to publish a revised edition of Goldenberg's *Gvozdev*.

In addition to those who assisted with and were named in the first edition; Peter Brown, Nelson Clark, Scott Hurlbert, Loren Leman, Jeannie Phillips, Jerry Phillips, Teresa Sowl, Shelly Stradling and Sharon Young, two additional individuals need to be mentioned with regard to the publication of this volume: J.L. Williams of Wales and Aaron Gray. Although I have never had the pleasure of meeting J.L. Williams, she was gracious enough to furnish the color photographs in this book. The technical assistance of Aaron Gray was particularly useful and appropriate in the publication of this volume. His life adds new meaning to the phrase "a friend indeed."

Anchorage, Alaska
June 13, 2001

AUTHOR'S PREFACE

In the history of Russian geographic discoveries and the study of Siberia, the Far East and Pacific Ocean in the 17th and 18th centuries there is a special place filled with the notable deeds of a geodesist named Mikhail Spiridonovich Gvozdev of the Peter the Great era. Under his command the Russian ship *Saint Gabriel* was the first European ship to approach the North American coast from the Pacific Ocean. This was the beginning of all the later discoveries in this region that led to the mapping of Alaska and the Aleutian Islands. Comparable to the First Kamchatka Expedition of Vitus Bering, it achieved considerable progress in the exploration of the northern most extension of the Pacific Ocean. Also connected with the name Gvozdev are other explorations of Eastern Siberia (Irkutsk Province), Kamchatka, the Kurile and Shantar Islands, Sakhalin, the Peninsula, and the Okhotsk and Bering Seas, all which are important in the development of our knowledge about the earth.

During the more than two hundred years since the discoveries by Gvozdev and his colleagues, not only has an intense scientific controversy developed, but an ideological struggle as well. Various groups of scientists have suppressed the importance of Russian research in the new region and misrepresented events of mapping the area. In addition, they suppressed the importance of Russian discoveries while promoting the significance of their own.

Gvozdev left us a valuable scientific heritage. Nevertheless, because of incomplete information, many questions remain unanswered in what has previously been written about his voyage. For example, arguments about who headed the expedition and Gvozdev's role in it remain unsettled. Though Gvozdev was a geodesist, his maps are still missing. He was commander of the ship but the original of his watch diary describing the voyage to Alaska cannot be located. He lived and worked in Siberia for more than thirty years but no one has previously written an accurate description of that period.

Historians of the 19th century mistakenly stated that Gvozdev entered Shliakhetskii Naval School in 1754 as an instructor (51, p. 402; 59, p. 85). Even contemporary publications mistakenly report that he was a teacher in this school in 1734 (30, p. 159; 29, p. 170; 27, p. 257). This school was not established until 19 years later. Also that Gvozdev was a student at the Slavie Latin School (25, p. 11; 52, p. 274). Following examinations in 1727, he reportedly received the rank of Warrant Officer (2, p. 36). This is also an error.

Biographies are usually written about celebrated representatives of science and culture, prominent statesmen and political figures, outstanding military commanders and other well-known personalities. Researchers usually study the extensively published literature. Then the diligent use of historical archives yielded new information. Russian historians owe a debt to the unassuming Gvozdev. The name Gvozdev is well known only in a small circle of scholars, even though his discoveries in the world of 18th century science were undoubtedly equivalent to those of Columbus, Vasco de Gama and Magellan. For a century and a half only a few reports and petitions written by the geodesist and only one brief, factually inaccurate, essay about him have been published (25). The arduous career of Mikhail Gvozdev is known only in the most general terms. The early period of his life, his leadership as head of the Kamchatka Detachment of the Shestakov-Pavlutski Expedition and the voyage itself lack thorough study.

Expansion of the documentary base helped fill in gaps in the life and achievements of the mapmaker and navigator. The basis of this study of M.S. Gvozdev is information available in the U.S.S.R and abroad, including material not previously known to exist that the author discovered in archives in Moscow and Leningrad (20). The several hundred documents containing the signature of Gvozdev and documents addressed to him now provide a clearer picture and a new perspective of the effort made by Gvozdev in the exploration of Northeast Asia. What is different is our view of the relationships between the major personalities of this historical voyage and their characters. Special attention is given to previously unanalyzed materials predating the voyage. Among these, preference is given to correspondence between the Assistant Navigator I. Fedorov, Navigator Ya. Gens and Gvozdev; the diary of Gens and instructions from Captain Pavlutski.

The setting for this study is the rise of a fledgling Russian scientific community during the early 18th century when Peter the Great initiated the mapping of Northeast Asia and the Pacific Ocean. Against this backdrop I have attempted an objective analysis of the scientific achievements of M.S. Gvozdev. Its results should contribute to a better understanding of the life and work of a geodesist and navigator during the time of Peter the Great. It will be seen that Gvozdev is a model of diligence, courage and determination, dedication to service and love of the Fatherland.

Table of Contents

Chapter		Page
	Preface	v
1	Mikhail Gvozdev: The Son of a Soldier	1
2	The Expedition to Northeast Asia	27
3	At the Head of a Sea Expedition to the Big Land in 1732	47
4	The Commander of the Kamchatka Detachment	73
5	The Cartographer of Port Okhotsk and of the Second Kamchatka Expedition	89
6	Retirement	111
7	The Pioneers of Alaska from the Pacific Ocean Side	117
	Dates in the Life of M.S. Gvozdev	131
	Documents by M.S. Gvozdev	133
	Bibliography	137
	Abbreviations	145
	Glossary	146
	History of the Names of the Islands in the Bering Strait	147
	Index of Names	163

Illustrations

Map of Northeast Asia by A.F. Shestakov, 1724	16
Early Maps of Northeast Asia	18
Illustration of *St. Gabriel*	26
Gvozdev's letter of May 1, 1732	50
Photograph of terminus of Cape Mountain	55
Photograph of Razorback Mountain	56
Photograph of Cape Mountain	57
Photograph of the three Diomede Islands	58
The Okhotsk Fleet of the Second Kamchatka Expedition	90
Gvozdev Map of 1743	121
Delisle's Sketch of 1738	122
Portion of Truskott-Müller Map of 1754-1758	123

Chapter 1
Mikhail Gvozdev: The Son of a Soldier

The complete absence of information on the genealogy and early life of Mikhail Spiridonovich Gvozdev has led to errors in previous research. In his petitions, Gvozdev writes about his study at the Moscow Academy from 1716 through 1718 and at the St. Petersburg Academy in 1719. Recalling that the first Moscow Academy was created by Peter the Great as an aristocratic educational institution and finding information about children of the nobility, the author looked to this documentary material for information about people of this class. Several files are dedicated to noble relatives of the Gvozdev family, including information from "the portfolios" of the historian of the Academy of Sciences, G.F. Müller. But I discovered nothing about Spiridon Gvozdev or his son Mikhail.

Later research in documents of the Admiralty College--the Admiralty Boards and their subordinate officers provided a starting point: a document prepared on July 9, 1753 for the establishment of the Naval Academy. It included a list of all the geodesists, thirty-four people, registered at that time with the Admiralty Board, together with a description of their social origin, major contributions, positions they occupied and last place of service. In the list of officers of geodesy, the section entitled "geodesists," by seniority, begins with the name "Mikhail Gvozdev of raznochintsy," [that is, intellectuals not of the gentry class]. It became clear why the name Gvozdev was not on the list of noblemen who participated in the Kamchatka Expedition, presented to the Admiralty Board in 1743 by M.P. Spanberg.[1]

At the time of Gvozdev's departure on the expedition in 1727, the Admiralty Board probably did not know very much about him. For example, one reported said his "condition was unknown." In another, "according to the rosters sent from the commanders of the Kamchatka Expedition regarding those on sea duty, report the appointment of Gvozdev to the Siberian region, serving as a member of the expedition."[2] Similarly, a short remark noted his service through 1728: "Appointed in 1716 to the Moscow School for the study of navigation, he went from this school to the St. Petersburg Academy in 1719 where he studied geodesy. In 1721, at the request of the Military Board, he served under Major General Volkov in the Novgorod District surveying areas for development of the yards of dragoon and infantry regiments. This assignment continued through 1725 when he transferred back to the Academy. A 1727 decree of the Admiralty Board sent Gvozdev to Siberia to do surveying and mapping work."[3]

There are two known petitions, or autobiographical sketches in existence: one dated August 24, 1743 and the other June 10, 1758. Both addressed the Empress. In the 1743 report, Gvozdev wrote the following about his work before 1728: "Appointed in 1716 by decree to the Moscow Academy of Science, I studied there until 1719. In that year I went to the St. Petersburg Academy to study geodesy. There I remained through 1721. In that year I went to Novgorod to the regiment commanded by Major General Mikhail Volkov surveying the areas of the regiment's headquarters. Stationed there until September 1725, I returned to the St. Petersburg Academy. In 1727, through examination (file 212) by Professor Pharvaron I became a geodesist. In that year, by order of the Governing Senate, I went to Siberia in the detachment of Captain Pavlutski and to the leader of the Cossacks, Shestakov."(25, p. 43).

The 1758 report reads: "Appointed to the Moscow Academy of Navigation Sciences in 1716, I remained there through 1718. I then went to the St. Petersburg Academy. In 1719, I was appointed as a student in the science of geodesy where I remained through 1721. In that year I went to Novgorod to the regiment under Major General Mikhail Yakovlevich Volkov surveying the rivers in the headquarters area of the dragoon and infantry regiments. I remained there through 1725 when I went back to the St. Petersburg Academy. Because of my knowledge of geodesy I earned the rank of geodesist. In 1727, by decree of the Supreme Secret Council and decision of the Admiralty Board, I was sent to the Governing Senate and from there to Tobolsk Province" (52, p. 157). The documents of the Admiralty Board agree with the information presented by the geodesist about himself. Differences exit only in minor details. But the passing mention of his social status (raznochintsy) required additional research.

Peter the Great established the special educational institution where Gvozdev studied, the Moscow Mathematical-Navigational School, in 1701. In the series of reforms that Peter initiated, there was a place for the preparation of Russian specialists in a newly born fleet and the regular army. Increased growth in productivity and the economic and political development of Russia at the beginning of the 18th century required immediate measures be taken to deal with backwardness in such divers fields as management, the economy, the army and navy, and culture. Success in foreign policy, development of industry, creation of governmental machinery, reform of the army and organization of the fleet demanded skilled personnel in all leadership positions.

The navigational school was primarily preparing navigators and shipbuilders, but its graduates worked successfully as engineers, artillerymen, geodesists and architects and in other environments especially related to mathematics and the physical sciences. Russian scholar Andrei Pharvarson, the astronomer and mathematician, headed the instruction of navigational sciences in this school. His assistants were English scholars Stephan Gvin and Richard Greis who were there at the invitation of Peter the Great. The talented Russian scientist and teacher Leonti Philippovich Magnitski, author of the mathematical encyclopedia of the day entitled *Arithmetic: the Numerical Science* (1703) headed the mathematics department.

The students there studied arithmetic, geometry, spherical trigonometry, geography, geodesy, drawing and technical drawing. They learned how to keep a ship's logbook and the basics of cartography and navigation. The high point of the curriculum was training in the theoretical fundamentals of navigational science and practice in navigation. The divisions of the navigation discipline were the "flat one" (navigation of short distances), "mercatorskaia" (navigation with the help of maps), and the "round one" (navigation of the arc of a great circle) (58, pp. 151-159). Practical studies at the astronomical observatory of Sukhareva Bashnia accompanied the theoretical classes. The students obtained the necessary skills by handling geodesic and astronomical tools (surveying compasses, astrolabes, sectors and quadrants).

The length of studies at the lower level of the mathematical school depended on the previous preparation and ability of the student. The soon to be famous hydrographer and cartographer F.I. Soimonov trained there for about three years beginning in 1708. Then L.P. Magnitski transferred him "to a school of foreigners, to Andrei Pharvarson and his colleagues," where by the end of 1711 he was appearing in the register "in flat navigation" (21, p. 24-25). After entering this school, geodesist Ignati Chicherin reached only the class of "flat navigation" by 1715. At the same time, others studied "the spheres" including later famous geodesists Kornei Borodavkin (at the school since 1710), Peter Chichagov (since 1710); Vasilii Yakovlev (since 1710), "in Mercator navigation," Tikhon Lodyshenski (since 1711), Arkhip Gerasimov (since 1710), "in flat navigation" Vasilii Leushinski (since 1712), and Nikita Sumarokov (since 1709).

In 1715 an instructor at the school designated those students who "were ready for practice." Among those designated for practical work were future geodesists Ivan Shekhonski, Ivan Khrushov and Fedor Molchanov.[4]

With the opening of the St. Petersburg Naval Academy in 1715, the functions of the Moscow Mathematical-Navigational School narrowed considerably. However, in private correspondence and even in official documents the Academy was often erroneously called "The Moscow Academy."[5] After Professor A. Pharvarson transferred to St. Petersburg the navigational department itself soon followed. 149 graduates from geometry and trigonometry transferred there along with Corporal I. Bykov in 1716.[6] Gradually the Moscow institution became more like a mathematical school. In 1727, 247 students were studying as follows: 131 students studied only arithmetic and 96, geometry; in 1732, two studied trigonometry, 20 studied geometry and 77, arithmetic (41, part III, pp. 417-419). By this time the structure of education at the Naval Academy assumed its permanent form. The system, established by decree of Peter the Great for training geodesists, existed until the development of the Navy Special Education

Corps. By 1727, 229 students studied at the Moscow Academy. Twenty-nine children of soldiers and 159 children of craftsmen studied at the Numerical School and Russian School. After graduation, the Admiralty Board appointed them to various positions. Students of the Academy trained in scientific disciplines.

Transfer from one class to another could occur at any time. The major criterion was the individual achievement of each student. The beginning class was "the science of arithmetic, "and the highest one was "big astronomy." Between these were classes in geometry, navigation, great circle navigation, creation of landscape maps and construction engineering. The 229 students were placed in the scientific classes as follows: three students from the Russian School who read and wrote Russian "were appointed to study arithmetic" with 58 other students; 30 students to geometry; and 108 students trained in navigation, artillery, fortification drawing and the foils of fencing. The distinguished students transferred to the next class. Four students who studied great circle navigation later received appointments to the navy as cadets and students of navigation. After training in mapping and construction engineering, five geodesists and 17 students of geodesy went to work in the provinces. Finally, four successful students of "big astronomy" received appointments as teachers and apprentices at the Academy (33, p. 51).

The instructors and staff at the Academy and the Moscow School were quite stable. A. Pharvarson headed them in St. Petersburg and L.P. Magnitski in Moscow. During forty years they trained hundreds of highly skilled specialists and developed a large number of textbooks and manuals. This educational achievement was the foundation for the establishment of a professional, secular, educational system. Unfortunately, this was not adequately appreciated in Russian publications. Whereas, in Moscow the instructional staff consisted of only one teacher, the Naval Academy staff consisted of nine people besides the professor. The latter, Fedor Alphimov, was assisted by five people in the discipline of navigation, two in the discipline of artillery, two in fortification, and one in geodesy. According to Admiralty Board records, from 1715 through

1739, Pharvarson was training students "in the sciences of arithmetic, geometry, trigonometry, spheres, parts of astronomy, geography, great circle navigation, geodesy, Euclidean elements, algebra, big astronomy, the theorems of Archimedes and other mathematical subjects. His salary was 960 rubles, 33 kopecks per year."[7] Several students of Pharvarson, who passed examinations under G.V. Kraft, H.N. Vinsgeim and I.G. Geinzius, professors at the Academy of Sciences, received appointments as instructors of mathematical and navigational sciences: Vasilii Ushakov and Aleksei Krivov in 1734, Avtamon Shishkov, Mikhail Chetverikov, Peter Kostiurin and Peter Biltsov in 1737, while Yukov Bukharin and Aleksei Isupov received appointments as assistants (apprentices).[8] Semen Saltanov, an instructor at the Academy since 1732, achieved "the rank of major" in 1740. On December 13, 1739 V. Ushakov was appointed to replace the deceased L.P. Magnitski at the Moscow School.

The number of the students at the school varied significantly. For example, in 1710 only 111 registered as students. From June 7, 1714 through February 1, 1715 447 became students. On January 26, 1716 the Military Office of the Czar ordered the Academy to have an enrollment of 300 students and an extra 30 students of geodesy. Regarding these students, Admiral F.M. Apraksin added, "Always have the prescribed number of children from privileged classes and supplement them with the children of the poor, sending any extras on to the Moscow School with orders and escorts" (41, P III, pp. 330-331). It would be difficult to formulate a more exact criteria for the selection of students. With those sent from St. Petersburg, the Mathematical-Navigational School had an enrollment of 498 in 1717, including 99 students without scholarships.[9] In 1719 the order was reaffirmed, requiring 300 children of the nobility in St. Petersburg and 500 in Moscow, while admitting students of all social classes. Children of the nobility owning five acres of land or fewer were to be sent from St. Petersburg to Moscow (41, P III, p. 350). In 1727, there were between 229 and 247 students. By 1732 there were 129 students at the Academy, including 22 in geodesy, 30 in flat navigation, 21 in geometry, and 56 in arithmetic.[10]

Some of the nobility opposed the required attendance at the School and Academy because their children wanted to avoid education by any means possible. The on-going inability to draw the required number of students led to the opening of these institutions to all children regardless of social standing. But there were certain age limitations (12 to 20 years old) and other regulations limiting those entering. As a rule, children from the Polish gentry joined the system much later than those from non-privileged social ranks. This circumstance may partially be explained by the fact that most privileged children received early education at home. According to our calculations, the average age of those joining the Academy from privileged classes was 17+ years old, while the average age of those from other social classes was 15+ years old.

In different years the ratio of privileged to non-privileged children varied considerably. In 1710, of 111 students, 11 were children of soldiers and 41 were children of nobility. Then in 1714-1715, of 447 students, 141 were children of soldiers and 116 were children of nobility. In 1717, of 498 students, 294 were from nobility, 11 from "priests' families," 113 of soldiers from the Preobrazhenski and Semenovski light infantry regiments, 10 from the merchant class, 23 from the landlord boyars, 4 of Polish origin, 4 from stable employees, 17 from the church ranks, and 22 children of soldiers from other military regiments."[11] Of the 22 students of geodesy in the Naval Academy in 1732, six were from privileged classes. In 1735, of 118 students at the Academy, 64 were from privileged and 54 from non-privileged classes. Of the 95 students at the Moscow school, 70 were studying arithmetic, 13 geometry, and 12 trigonometry: 24 belonged to privileged and 71 to non-privileged classes (68, pp. 56-57).

On December 24, 1715 by a decree of the Czar: "The son of a soldier from the Semenovski light infantry regiment, Mikhail Gvozdev is to be sent to the Moscow Mathematical-Navigational School by the Preobrazhenski Department. While training he and his friends were to go to Fedor Urevich Romodanovski at the nearest Preobrazhenski Department. He was a stolnik, a person who looked after the Czar's food service. The decree noted that

Mikhail would not have been sent anywhere without a decree"[12] nor was he "granted a salary from the Emperor." On obtaining the Czar's order, the President of the Admiralty, F.M. Apraksin, gave the following direction to the Admiralty office on December 31, 1715 "requesting in the book" that the young man go to school: "and under this remark the above described soldier's son Mikhail Gvozdev was sent to school on January 2, 1716."[13]

The Preobrazhenski Department and its head, F.U. Romodanovski, had a horrible reputation. Peter the Great created the Preobrazhenski Department in 1686 in a village near Moscow called Preobrazhenski. Its purpose was to manage a regiment of boy-soldiers, which later became the Preobrazhenski and Semenovski regiments. The Czar used them in his struggle for power against Czarina Sophia. With broadened functions, the department received authority to conduct investigations of political crimes and was in charge of the Moscow police as well as the two infantry regiments. Under the personal leadership of Peter the Great, the Czar directed the activity of the Department as an internal state institution of Russia against enemies of Peter's reforms and used it to suppress revolts against serfdom. Even the mere mention of the torture chambers of the Preobrazhenski Department inspired horror. As a participant in the childhood military games and maneuvers of the Czar, F.U. Romodanovski was very loyal to Peter the Great and had his complete trust. Armed with enormous power, Romodanovski distinguished himself by his unusual cruelty in conducting investigations and searches.

The requests of the students of the Navigational School are preserved. One of them, on monogrammed paper, is presented in its entirety as the earliest document in existence containing the signature of Gvozdev: "Supremely powerful Czar, Merciful Ruler! By your great decree, we your undersigned servants are studying at the School of Mathematical-Navigational Sciences. Your imperial scholarship, granted to us, your servants, does not provide to us, as orphans, any financial assistance from either the Prebrazhenski Department. Most merciful Emperor, we are asking your Majesty to provide a stipend of your determination for us for Majesty's servants of the Mathematical School, students Fedor Apushkin and Mikhail Gvozdev. On January 30, 1716, to this petition, Fedor Apushkin put his hand. To this petition, Mikhail Gvozdev put his hand."[14] Admiralty Commissioner A.A. Beliaev prepared an analysis and submitted it with a summary of the request to serve as the basis for a response. He decided to provide the petitioners with 4 denga, 2 kopecks per day, or 20 altyn (1 ruble, 6 altyn, 4 denga) per month. Based upon F.M. Apraskin's order of April 19, 1712 financial aid of 4 denga per day was given to each of the 65 children of soldiers from the Semenovski and Preobrazhenski regiments. As they "advanced academically, their allowance increased to 6 denga per day and nothing more."[15]

On February 6, 1716 the Admiralty Department decided to pay four denga per person per day to students of arithmetic F. Apushkin and M. Gvozdev, children of soldiers of the Semenovski Regiment. It was decided to advance funds beginning February 16 and inform schoolteacher L.F. Magnitski by decree. Under this decree, there is a second signature of Gvozdev: "And by this mark Mikhail Gvozdev has taken the decree directed to the teacher and has signed it."[16] Once the stipend of the members of the arithmetic class was determined, the students received 4 denga per day with the prospect that it would increase according to their achievement.

Of the 498 students at the Navigational School in 1717, M.S. Gvozdev, together with future geodesists Semen Chichagov and Aleksei Thikhmanov, were among 113 soldier's children of the light infantry regiments of Preobrazhenski and Semenovski.[17]

Based on the earliest documented information about Gvozdev, it is possible to say with certainty that Gvozdev, like others of common birth, did not belong to the privileged gentry. Examples of these include M.V. Lomonosov and other children of soldiers who became geodesists such as A.D. Krasilnikov, P.I. Chichagov, T. Lodyshenshi, V. Yakovlev and N. Sumarokov. It is well known that Peter the Great was a man of immense energy and activity and often brought into the service of the State "people of various social ranks" to reinforce the government machinery. M.S. Gvozdev was the orphaned son of a soldier of the Semenovski Infantry Regiment. His father was probably killed in one of the battles of the Great Northern War. He was raised in the regiment without any financial aid. When he reached a certain age, he was sent on to the Navigational School by a decree of the Preobrazhenski Department. What is important is that by this decree his future was made secure. After his training he was to return to this Department under the authority of F.U. Romodanovski. He could not be transferred anywhere without a special order from the Czar, apparently because of the special dispensation of Peter the Great on behalf of the officers and men of the Semenovski and Preobrazhenski infantry regiments, a circumstance noted by various historians.

It is difficult to pinpoint the year of Gvozdev's birth for lack of proper historical records. Only through his classmates' ages which are known can we suggest that he was no older than 14 to 16 years at the time of his registration as a student in January 1716. Taking into consideration his status as an orphan, one cannot ignore the possibility that he entered school as early as his 12th year. Accordingly we may suppose that M.S. Gvozdev was born at the beginning of the 18th century, but no later than 1704. According to other less satisfactory calculations, it would seem that by the time of his final qualifying examination for geodesist in 1727, Gvozdev was between 28 and 33 years old and that therefore his birth was sometime between 1694 and 1699 (37, p. 142).

Gvozdev's resumes do not mention the particular sciences he studied. Most likely they were the same disciplines as those recalled by Peter Skobeltsyn and Ivan Svistunov, his contemporaries at the Academy. After mastering arithmetic, geometry, trigonometry, flat and Merator's navigation, and passing Professor Pharvarson's examination, the school of Mathematical and Navigational Sciences granted joining freshmen their first degree, the title of student of geodesy. The prerequisite for a second degree, the title of geodesist, was the study of geodesy, drawing and cartography.[18] I.S. Svistunov studied these subjects for five years and P.N. Skobeltsyn for eight. Generally, the length of the studies at the school and Academy was usually no longer than five to eight years. Those studying five years were I.F. Khanykov, D.A. Mordvinov, A. Tolubeev, F.A. Zubov, Ivan and Leonri Isakovs, and

A.D. Krasilnikov; six years were F.I. Teglev and E.F. Safonov; seven years were M.S. Gvozdev and S.A. Arsenev; and eight years were V.M. Zurov, M. Ignatev and P.N. Skobeltsyn. Sometimes work assignments disrupted studies as in the case of F.F. Kuchin, whose work lasted from 1708 to 1721, or with Gvozdev himself.

In November 1726, the Admiralty Board approved a plan, presented by A. Pharvarson and F.D. Alphimov, suggesting a time schedule for studying the various disciplines at the Moscow Academy. Generally, the length of study was six years and nine months. "One year for the study of arithmetic, eight months for geometry, three months for trigonometry, three months for flat navigation, five months for Mercator's navigation, one month for diurnal, three months for spherical trigonometry, four months for spherical astronomy, one month for geography, one month for round navigation, one year for geodesy, and one year for painting and fencing" (67, p. 109).

Geodesists and students of geodesy carried out various tasks in the techniques of mapping, measuring soil, marking borders, constructing defensive lines, building roads and canals, etc. Usually, by request of a board, especially one of a military, estate, foreign affairs, palace office or other centralized state institution, the Senate would, by special decree, direct the Admiralty Board to provide logistical support for special tasks using the geodesists from the Naval Academy. Examples of such assignments to various regions would be the dispatch of navigator F.U. Molchanov to Siberia in 1717; Mikhail Ignatev and Fedor Baluev sent abroad with the diplomatic delegation of L. Izamailov in 1719; geodesists Yakov Vedrin, Konstantin Zinovev, Daniel Pokhvisnev and Stepan Urenev to Kronstadt "on the canal;" students of geodesy Yakov Radishev, Semen Lykov, Gregory Kudrin, Aphanasi Myshetski, Nikon Kostochkin and Isaac Pustyshkin "to the construction of the Great Ladozheski Canal"; and Ivan Balashov "for the construction of a perspective road" in 1721. The most significant and simultaneous departure of geodesists and students "for the description and development of land maps" in the regions and provinces took place at the beginning of 1721. It was then that the Czar's plan for the first nationwide instrument survey of Russia began (50, Volume 6, No. 3682, 3695).

On March 14, 1721 the following geodesists were sent to the provinces to survey the land and make maps at a salary of 6 rubles per month paid by the Provinces: Tikhon Lodyzhenski and Stepan Ignatev were sent to the Moscow Province, Ivan Khrushov and Boris Baturin to Kievskai Province, Isai Krapivin and Stepan Kuchin to Kazanskai Province, Ivan Shekhonski and Stepan Orlikov to Nizhnegorodskai Province, Arkhip Gerasimov and Yakov Philisov to Arkhangelskai Province, Fedor Kuchin and Vasilii Yakovlev to Smolenskai Province, and others to various provinces and regions. Usually two geodesists were sent as a team. According to explanations of historians of mapmaking, under these conditions one of the cartographers was junior, usually a student of geodesy who worked as an assistant (65, p. 83). By July 1727, 44 geodesists and 26 students of geodesy worked on such projects. In January 1732, registration totaled 111 persons in such work (68, p. 48).

In 1718, after three years of training at the Navigational School, M.S. Gvozdev transferred to the St. Petersburg Academy where he continued his education through 1721. We might suppose that in St. Petersburg he was receiving the same salary as the rest of the students. As the result of a decision in 1719 by G.G. Skorniakov-Pisarev, head of the Naval Academy, the students' salaries increased as they advanced in their classes. For example, 1 ruble, 45 kopecks per month in the class of geometry, 2 rubles, 13 kopecks in the class of round navigation, 2 rubles, 88 kopecks in the class of flat navigation, and 2 rubles, 88 kopecks in the class of geodesy (67, p. 107).

Upon graduation as a student of geodesy in 1721, Gvozdev received his first professional assignment. At the request of the Military Board, Gvozdev worked as part of the detachment of Major General M.Y. Volkov, whose responsibility it was to audit the results of the population census taken in the province of St. Petersburg in 1719-1721. His office for the census was in Novgorod. The instructions published in a decree dated March 1722 governed the conduct of the work. Additional duties included such activities as confiscation of land illegally appropriated by serfs, investigation of local conflicts and laying-out military subdivisions.

In a report[19] from Novgorod, dated September 21, 1721 M.Y. Volkov informed the Senate about the initial stages of his work, and in particular about the instructions in the order of August 30, about the registration of the Dragoon and Infantry Regiments. He referred specifically to the Senate decree of March 5, 1721 which read in part, "It is ordered that in the Novgorod District, all men in the military Dragoon and Infantry Regiments are to register and to pay a tax of 32 altyn and 2 denga per person...."[20]

After M.S. Gvozdev worked at this assignment for about four years, he received an assignment to describe the rivers and other features in the Novgorod District for "the settlement" and building of the regiment headquarters. In September 1725, he returned to the Naval Academy to resume his studies. Here, after successfully passing the examination of A. Pharvarson, "based upon his knowledge" of the sciences and professional decorum, Gvozdev received the title of Geodesist in 1727.

Chief Secretary of the Senate, I.K. Kirilov headed the department of Russia's cartography and systematically inspected the work of the geodesists. He received urgent orders from the Moscow Senate to forward certain information about some of the work of the geodesists. He was to find out "if they were taking their leisure" and "what work they were involved in." They wanted to know specifically about geodesists A. Sipiagin, S. Kashintsov, F.K. Ezhevski, K. Borodavkin, and N. Sumarokov. Also, "why was the development of the Moscow plan taking so long" (the so-called Michurinski plan of Moscow, published in 1739). Also, where were A. Ia. Krotkov and F.D. Lavrov, who earlier were sent to survey the land of the palace village Sophino. As directed, he checked the report of the Admiralty Board about the dispatch of I. Balashov to Pochep to survey land with Colonel Davydov. In 1726 he ordered the dispatch of Fedor Baluev and Mikhail Gvozdev to the Palace Office to survey the Palace Estates, "if employed in this work or free." The Senate office, which had prepared a report of detailed information about the work and location of each geodesist, forwarded the inquiry about Gvozdev to the Palace Office.

As it turned out, the information the Admiralty Board had about Gvozdev was in error. According to the records of the Palace Office in Voronezhskai Province and the Tambov palace villages, three geodesists (I. Balashev, F. Baluev and M. Stromilin) were surveying land and preparing maps. Two of them returned to Moscow so that "they would not live in leisure and take their salaries in vain." As for geodesist Mikhail Gvozdev, he was not sent to the Palace Office.[21] Accordingly, the report about sending M. Gvozdev to survey the palace land in 1726 was the result of a bureaucratic error.

While Gvozdev was finishing his studies at the Naval Academy in 1724-1727, high state officials carefully examined and approved the organization of an expedition to the Yakutsk area under the leadership of Afanasi Fedotovich Shestakov. His knowledge of hardships, local conditions and all the details of service in the Russian garrison located in Irkutsk enabled Shestakov to propose many measures to stabilize relations with the local population, "search for new land," explore land, river and sea communications, regulate the activities of servitors and reinforce Russian ostrogs. The exact regions of exploration were the Okhotsk Sea, the islands off the northern coasts of Asia, "the lands opposite the Anadyr Cape," and the Kurile Islands. In this way, A.F. Shestakov emphasized the disparate character of the expedition from that of the First Kamchatka Expedition organized in St. Petersburg in 1725.

The plans of the leader of the Yakutsk Cossacks were warmly received since they coincided with the political expectations of those surrounding Peter the Great and Catherine the First. They also reflected the aspirations of the imperial empire to expand Russian territories, to speed up the movement to the northeast, to obtain additional sources of state revenue and to study new possibilities for trade.[22] Supposedly, Vitus Bering knew about Shestakov's expedition. When both expeditions were underway they were to help one other. The thinking of A.A. Pokrovski, a specialist of early texts, is mentioned here. He explained that because of the uncertainties about geography in those days, almost simultaneously and through the initiative of the same institutions, two similar (he thought) expeditions set out: one under Bering and one under Shestakov. He thinks the reason for this strange dual effort was the map of Shestakov with its representation of "The Big Land" opposite the mouth of the Kolyma River. Accordingly, he concluded that one expedition, by Bering, was to explore the East Sea, while the other, by Shestakov, was to explore seas to the north (69, pp. 12-13).

Such a conclusion contradicts the well-known tasks entrusted to the expeditions. For Shestakov, the exploration of the northern coast of Asia and adjoining islands turned out to be of secondary importance. His main emphasis was in Chukotka, Kamchatka and the Okhotsk and Bering Seas. Here he was to concentrate on non-scientific problems: putting tax collection in order, reconstruction of old ostrogs and building new ostrogs and winter cabins.

Shestakov's success depended largely upon having accurate geographical information about the Yakutsk - Kolyma - Chukotka - Kamchatka region including information about the natural habitat of the indigenous peoples, about the means of communications and about the availability and number of government personnel. During the discussions of the project Shestakov used "the special map" which, despite its cartographic naiveté, is a valuable historical source because it reflected a definite vision of the surrounding world. Together with a series of other Yakutsk working drawings of earlier decades, this map reflects the ancient Russian cartographic tradition. The understanding of such maps show their practical uses and, regardless of imperfections, they were generally reliable (19, pp. 137-160). In a simple, clear way Shestakov expressed the concepts of Yakutsk servitors. In 1722, Siberian Governor A.M. Cherkaski reported about them and "the navigator requested permission to search for new lands and islands" using them.[23]

Maps presented here by Shestakov and Kirilov with their ideas of "Kamchadalia" and Anadyr are working drawings that need no special explanation. The differences between them are especially noticeable because of the different basic purpose of each map. The influence of the sources used by Kirilov is noticeable only on his map of 1724. The Russian maps of "Kamchadalia" in the atlas of J.B. Homann omit the contours of the vast land opposite

Kamchatka, the islands in the Lamskoe Sea, the two islands and the unlimited land (Alaska) opposite the Chukotka Peninsula. On all the maps, the depictions and the locations of the Kurile Islands are different. All these cartographic materials served as reference manuals for completing the tasks set before the Bering and Shestakov expeditions. Dissimilar representations on these maps of the northern coast of Asia, Chukotka and Kamchatka, the Kurile ridge, and the islands in the Okhotsk and Bering Seas were often explained by the competing ideas that served as the basis for the different working drawings. The conceptions of Kolymsk, Anadyr, Kamchatka and Yakutsk peoples of Northeast Asia varied significantly.

Map of Northeast Asia by A.F. Shestakov, 1724.

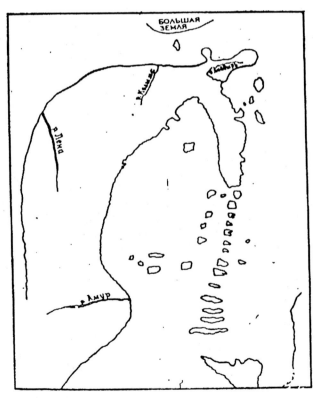

Shestakov's project did not contain any notes of a scientific nature. The question of sending a geodesist was not addressed. The emphasis of sending of a specialist to compile maps and plans of the land and sea regions was contained only in the directions of March 23 and May 3, 1727. On April 11, the Admiralty Board forwarded to the Senate the decision to send a geodesist. At the beginning of May, immediately after the massive departure of geodesists from the Naval Academy to work in the provinces, the selection of Mikhail Spiridonovich Gvozdev was made. A highly qualified specialist, he was a well-trained geodesist; a cartographer professionally prepared for surveying and compiling of maps of both land and sea. In school and at the Academy, Gvozdev had acquired the habits of a combat officer, received sea training, learned the fundamentals of ship piloting, ship logging, use of survey and navigational instruments; and, if in an emergency, he could accomplish the duties of a ship's navigator as well.

On May 20, 1727 M.S. Gvozdev was "taken" from the Academy. The same day the Admiralty office informed the Board that M.S. Gvozdev had not received a salary for the year 1727 and requested, because of his being dispatched, payment of a year's wage at the expense of the Admiralty with the deduction of a month's salary because of the advance payment. The Board decided to pay him from the St. Petersburg treasury 66 rubles for the year 1727 (72 rubles was a year's salary, less 6 rubles for advance payment).[24]

On June 9, the Navigational Office sent to the Senate all "the Admiralty servitors" with their appropriate tools. Appointed to the expedition of Shestakov were the geodesist, ten sailors, assistant navigator Ivan Fedorov and the Dutch navigator in Russian service, Yakov (Jacob) Gens. The sailors received supplies and rations for the first half of 1727 and cash through 1728 based upon the following calculations of monthly salary: some flour and cereal, and 2 pounds of salt, 1 ruble, 50 kopecks, "with a deduction for uniforms" of 30 altyn and for the sailors of the second class, 1 ruble and 20 altyn." A navigator's rate of pay was 12 rubles per month and for "the additional" thirteenth month. An assistant navigator was paid 7 rubles. For their work, both also received navigational supplies: 2 pel compasses; 8 ordinary compasses; two half-hour, one minute and one half-minute bells, six destei of Alexandria and fine quality paper, three gross of writing paper and two telescopes.²⁵

M.S. Gvozdev received an iron chain for measuring distances, a theodolite and azimuth compass for measuring compass points of a log line and a quadrant for determining the astronomical latitude by the sun or the North Star. Also issued were instruments used for laboratory work including a case of drawing instruments, a pair of compasses, "big vernier calipers," a simple compass with three legs, a ruler with division signs, and Mercator's and other tables."²⁶

Maps of Northeast Asia.

a) the map of "Kamchadalia" 1722 (from the atlas of J.B. Homann 1725)

b) the map of I.K. Kirilov 1724

в) "Anadyr drawing" not earlier than 1727-1728

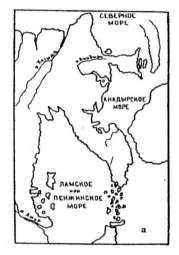

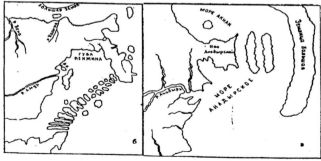

A.F. Shestakov thoroughly examined the future participants of the expedition. He repeatedly expressed his concerns to the Senate about some of the men assigned to him. While there was not one word mentioned about the young geodesist, he gave definite opinions about the seamen. Thus on June 16 the persistent Cossack leader considering the expeditions requirements for sailors of great endurance, experience and health reported the following; "And those sailors sent by the Senate are not knowledgeable about the local way of life. I am concerned that they might desert the expediton."[27]

His evaluation of the physical condition of Ia. Gens and I. Fedorov was no less strict: "About the navigator and assistant navigator, they should not be on this expedition at all, because the navigator lacks stamina and is not very skilled in Russian. The assistant navigator is ill. Apart from them, we may be able to find people familiar with local sea conditions in the city of Yakutsk."[28] Historian A. Sgibnev, who had a negative attitude towards Shestakov, wrote that the Cossack leader hated educated people and tried to get rid of them. It is hard to disagree with this position. But the fears of Shestakov were well founded. He knew the severity of the conditions of life and work in the northeast very well. He was trying to obtain not only knowledgeable people for the expedition, but also those of strong body and spirit. In July 1732, before departing on his second expedition, Bering recalled his meeting with Shestakov's navigator who was ill. Bering said he didn't think he would survive.[29] Looking ahead, we may note that neither Gens nor Fedorov survived the hardships.

By the decree of June 19, 1727 the Senate ordered Shestakov's expedition to leave St. Petersburg for Siberia immediately. But the next day, both the navigator and geodesist insisted upon appropriate outfitting of equipment and tools. For "the sea voyage" Ia. Gens requested two additional compasses of seven and nine inches, six destei of Alexandria paper, seven destei of fine quality paper, 1.5 stopa of writing paper, a piece of magnet for repairing compasses, one 5'X35'X2' chest for storing materials, two telescopes and two sextants. Gvozdev also sent the Senate a report citing the impossibility of quickly accomplishing his assigned tasks and the need for additional provisions. He noted that he was sending him to Siberia "to describe and create maps of the Eastern and Northern seas and some islands." Describing these seas and islands alone would not be accomplished quickly. Geodesists, he noted, are usually sent to such places in pairs and fours. They were sending only two students of geodesy to help him.[30] His observations were well founded.

While making the request for additional help for this expedition into remote regions, he was well aware of the methods of surveying and cartography. Gvozdev knew very well that the Navigational Ministry usually sent graduates of Moscow and St. Petersburg from 1718 through 1726 in pairs and fours "for describing places and creating land maps." The Board's files of January 2, 1725 and July 1727 mark the dispatch of the Czar's geodesists and students of the geodesy (68, p. 46-47). The following people were sent from the Board to the Senate for work: to northern provinces, on May 31, 1718 F.F. Luzhin and I.M. Evreinov; on February 4, 1719 I. Zaharov, and P.I. Chichagov, on February 19, 1724 geodesist P.N. Skobeltsyn and students I.S. Svistunov, D. Baskakov, and V.D. Shetilov to Astrakhanskai Province; on September 19, 1720 I. Chicherin, A. Tolubeev, A.A. Sipiagin, and G. Makarov to the Nizovaia Expedition; on March 15, 1722 S. Kashintsov, F.I. Teglev, G. Okulov and I. Lebedev to Voronezhshai Province; in 1718, K. Borodavkin and N. Sumrokov to Moscow Province; and V. Surovtsov, C. Dyakov and others in 1722 "for the registration of men."[31]

The expectations of Gvozdev were well described by an unknown author in a drawing in the first half of the 18th century. On this map the line scale, in minutes and Russian versts, is enclosed in a frame. Two cartographers hold the ends. In Gvozdev's report of June 20, 1727 he also asked for the materials necessary for such work. The extensive list attached is very interesting in the history of cartography. Gvozdev, who had already practiced his profession, prepared the detailed list characterizing the needs of a geodesist and the various materials used in making maps. With the background of stories of geodesists who had worked in Siberia, he correctly foresaw every eventuality of the extreme difficulties of the remote area.

Gvozdev insisted upon being provided with the following:

Alexandria paper	4 destei	guter marin	8 zolotnik
Leather camer-obskur w/glasses		thick pencils	1 dozen
writing paper	4 stopa	thin pencils	3 dozen
colors:		shade brushes	2 dozen
bakana turetshai	24 zolotnik	drawing brushes	1 dozen
bakana venetsianskaia (crimson)	24 zolotnik	glue/white wash	48 zolotnik
		sealing wax	1 pound
cinnabar	32 zolotnik	gum	1 pound
"yar" venetsianskaia (bright green)	1 pound	alum	24 zolotnik
		nut gal	1 pound
golubets	20 zolotnik	sal ammonia	
gomgegut (yellow)	48 zolotnik	Indian Ink	
purple	7 zolotnik	Chinese Ink	2 boxes
cormin rota	10 zolotnik	drakoglut	12 zolotnik
krutic (blue)	48 zolotnik	light zhezhgel	24 zolotnik
dark zheshgel	36 zolotnik	sea telescope	1
berline light blue	15 zolotnik		
half-Alexandriapaper	5 destei		

In addition, the list included "a thin slab for grinding colors," and four boxes for packing the quadrant, astrolabe, colors, paper and other materials.[32]

On June 28, the Senate discussed the reasons for the delay of the departure of Shestakov's expedition. On July 14, a decision was made: forward to the Admiralty Board a decree to send two apprentices of geodesy to assist Gvozdev; and "obtain from Gvozdev and the navigator a list of the supplies and materials needed and the prices thereof." The next day Gvozdev appeared before the Senate where he provided a detailed explanation of the additional materials he needed. The original of this "fairytale" is preserved. He made quantitative changes from his first request and substantiated the purpose of the supplies. It is necessary to have Alexandria paper and half-Alexandria paper to make land maps and exact descriptions of the islands. The book paper, writing paper and fine quality paper are needed for drawing rough drafts, making notes in registers and journals and piecing maps together. The colors are for drawing the land maps and islands and the dividing and marking the lands.

The enumeration is preserved with some small changes: purple, 3; cormin rotri, 4; berline blue, 7; guter marin, 4; and drakoglut, 6 zolotnik. The amount of sealing wax required was reduced to 15 zolotnik. The Chinese ink and pencils are for drawing maps. Brushes are needed for shading and painting colors and ink. The telescope is "for direction at sea and for viewing islands at long distances to see distinguishing characteristics." The camer-obskuris for better land surveying of prospective places and of inaccessible ones. The thin slab is for grinding colors with a courant. About the exact prices are for the materials, it was impossible for Gvozdev to say. In this manner, he concluded his testimony to the Senate, attesting to it with his signature.[33] A Senate extract of the requirements of geodesist ensigns Vasili Somov and Peter Lupandin, who had were dispatched in 1740 to the RussianTurkish border provides an idea of the prices of some of the geodesy instruments and materials in the first half of the 18th century. It is important to keep in mind that the prices of these supplies changed over the years.

The certificates of the Academy of Science, the Naval Academy and the Defense Office for prices of instruments and materials purchased by Lupandin and Somov serve as a basis for compiling a general estimate of the expenses. In them the following prices are indicated:[34]

Astrolabe	20 to 30 rubles
quadrant	50 rubles
sazhen chain	7 rubles
1' to 2' telescope	2 to 20 rubles
1 destei Alexandria paper of "the French" kind	1 ruble, 40 kopecks
4 stopa book paper	8 rubles
1 stopa "every day paper"	1 ruble, 20 kopecks
1 box of Chinese ink	2 rubles

12 white wood pencils	1 ruble
12 brushes	5 kopecks
paper glue	5 kopecks
1 lb "yar" venetsianshaia	2 rubles, 40 kopecks
48 zolotnik of "gamigut"	60 kopecks
1 zolotnik of carmine	2 rubles, 50 kopecks
1 zolotnik of ultramarine	5 rubles
1/4 lb Berlin azure	3 rubles, 84 kopecks
100 goose pens	60 kopecks
48 zolotnik of Komed	10 kopecks

On July 17 the Senate forwarded Gvozdev's "fairytale" to the Admiralty Board together with the request to send two geodesists and to add the listed instruments and materials. Nevertheless, on July 24, the Board informed the Senate that in its opinion, no more than two apprentices should be sent with Gvozdev. "The geodesist should carry out the prescribed task by himself." Provide him with some of the dispatched servitors for help according to the number of people customary. The number of apprentices of geodesy at the Academy was insufficient. If there was a need in another area, there should be some one to dispatch. The Board also declined the request for the supply of materials for lack of funds. The Board recommended that the Senate purchase all the necessary items at the trade market. The Senate took the arguments of the Admiralty Board into consideration. On July 31, the Senate forwarded the July 28 decree to the Board instructing it to purchase the prescribed instruments and materials in St. Petersburg or in Moscow and to give them to the geodesist Gvozdev immediately, and to report at once, with a receipt for the purchases."[35]

Since that time Gvozdev's name became permanent in the lists of requisitions prepared by the Senate and the Admiralty Board. In the July, 1727 register of "the geodesists sent to various areas," marked him as sent "to the High Senate to be dispatched to Siberia for the development of land maps of the sea and other local areas" (68, pp. 48, 50). In the March, 1732 register, the year of the dispatch is inaccurately stated as 1726. The Senate chronicle on March 8, 1733 notes: "and for the dispatching of Mikhail Gvozdev to Kamchatka with Shestakov." On March 22, 1734 the Register of the Board marked the location of the geodesist as in the Kamchatka Expedition. The same is witnessed on the roster of the same year in the notes of I.K. Kirilov, "Mikhail Gvozdev at Kamchatka with Shestakov." On the roster on November 10, 1735, "Mikhail Gvozdev is in Kamchatka." The rosters of 1735 and 1736 starting the last names of the geodesists, taken by the Senate for various purposes, include "in 1727, Mikhail Gvozdev."[36] Gvozdev himself, in his autobiography of 1743, wrote about his dispatch by the Senate in 1727 to Siberia with the detachment of A.F Shestakov and D.I. Pavlutski and of his staying in Kamchatka through 1735.

Footnotes for Chapter One

1 TsGAVMF, f. 216, Case 99, p. 374.
2 Ibid., f. 212, Case 27, p. 121.
3 Ibid.
4 Ibid., f. 176, Case 106. pp. 1344-1347; Case 115, p. 1069; f. 233, Case 104, p. 213.
5 Ibid., f. 248, Bk 1072, p, 136, AAN, f. 3 # 1, Case 2332, p. 34.
6 Ibid., f. 176, Case 115, p. 1125 and 1199.
7 TsGADA, f. 248, Bk 1205, p. 444.
8 Ibid., pp. 335-336, Bk 794, p. 201.
9 TsGAVMF, f. 233, Case 149, p. 85.
10 TsGAVMF, f. 21, Book 52, pp. 1-5.
11 TsGAVMF, f. 233, Case 149, pp. 82-84.
12 Ibid.
13 Ibid.
14 Ibid., p. 247.
15 Ibid., pp. 247-248.
16 Ibid., p. 249.
17 Ibid., p. 82.
18 TsGAVMF, f. 216, Case 52, p. 485; f. 248, Bk 287.
19 Donoshenie: Document of a subordinate to his superior.
20 TsGAVMF, f. 248, Bk 690, p. 1.
21 Ibid., Bk 528, p. 200.
22 Ibid., Bk 690, p. 249, also Vol. VII, 50, No. 5049; p. 56.

23 Ibid., p. 275.
24 Ibid., p. 294.
25 TsGAVMF, f. 216, Case 5, p. 96.
26 TsGAVMF, f. 248, Bk 690, pp. 298, 306.
27 Ibid., pp. 300-301.
28 Ibid., p. 301.
29 Ibid., Bk 664, p. 102.
30 Ibid., Bk 690, p. 308.
31 TsGADA, f. 212, Bk 29, pp. 364-366
32 TsGAVMF, f. 248, Bk 690, pp. 307-308
33 Ibid., pp. 309-311.
34 Ibid., Bk 1072, pp. 359-361.
35 Ibid., Bk 690, pp. 312 and the other side of p. 312.
36 TsGADA, f. 248, Bk 779, p. 58; Bk 664, pp. 177, 326, 388, 402, 410; Bk 716, p. 379.

The *St.Gabriel*

Chapter 2
The Expedition to Northeast Asia

With the purchase of goods, instruments and materials in Moscow, Shestakov's expedition departed for Northeast Asia during the first days of August 1727. Craftsman Ivan Speshnev joined the expedition September 6.

Beginning with historians G.F. Müller and A.S. Sgibnev, all the research usually identifies A.F. Shestakov as head of the expedition, appointed by the Senate to be the chief commander of the whole northeast region of Siberia. But this is not true. None of the decrees make such a statement. In reality, Shestakov headed the expedition on his own only during its preparations in St. Petersburg and Moscow and on the way to Tobolsk. In the capital of Siberia the governor appointed the Captain of the Siberian Dragoon Regiment, Dmitri Ivanovich Pavlutski to head the expedition "mutually" with Shestakov. He commissioned the making of decisions by "mutual agreement,"[1] undermining from the very beginning the wise principle of having one person in charge. But the masterful and proud Yakutsk leader of the Cossacks often interpreted government decrees in his own way.

The manning and outfitting of the expedition to the East took more than two months. In the fall of 1727 Gvozdev and his detachment arrived in Tobolsk with 41 men. The size of the expedition grew as it advanced. In Eniseisk Province, 83 joined. In Irkutsk, 30 more were added. Then in Yakutsk, 426 people joined the expedition.[2] From the ostrogs of Kirenski and Cheruiski and the Krivolutski settlement, Shestakov selected 94 Ilimski peasants. He presumed that "without such selected peasants' children on the team" the expedition would be impossible because of their use in such tasks as making skis and dog sleds and preparing trails along the way. They can hunt any beast and will support the new recruits in the expedition.[3] Gens and Fedorov received additional navigators' supplies from the Tobolsk provincial office, including 150 arshins of canvas, 160 pounds of sailing thread and 50 sewing needles.[4]

In Tobolsk, Gvozdev became acquainted with Monk Ignati (Ivan Petrovich) Kozyrevsk, the explorer of the Kurile Islands and Kamchatka. The Siberian governor offered Kozyrevsk to the expedition "for explaining Kamchatka, the northern part of the new land and the islands and people of which he was knowledgeable."[5]

On November 28, 1727 the expedition headed for Eniseisk from Tobolsk. Then overcoming the Argarski and Leno Angarski mountain ridges, they arrived on the Lensko Angarsk plateau. Here, to the south of Ust-Kut, they built ships and started to sail along the Lena River to Yakutsk. The ship *Evers* was built under the direction of Kozyrevsk. It provided transportation for 280 servitors from the Ilimski district to Yakutsk. Shestakov built a smaller ship at his own expense, which on May 9, 1728 having moved along Aldan, Ma and Yudoma Rivers, reached the East Sea. Twelve large and small rafts were also constructed.

First the Tobolsk regional office and then the Senate became concerned about tense relations between Shestakov and Pavlutski. On April 26, 1729 Vice-Governor, I. Boltin, prepared for I.K. Kirilov, a summary of the reports of both leaders testifying "about the unethical actions of the other."[6] There was no agreement between the leaders of the expedition. The sailors and Kozyrevsk surrounded Shestakov. On the way, the Cossack leader brought sailor L. Petrov under his wing and made him his direct assistant. Gens and Speshnev sided with Pavlutski. There is no information about the positions taken by Fedorov and Gvozdev, though the geodesist, according to his autobiography in 1758, "and was appointed to the detachment of the Dragoon Regiment of Captain Pavlutski to the Anadyr crew" in Tobolsk (52, p. 156).

Of course the discord and quarrels between the commanders, affected the accomplishment of the various tasks over time. On receiving the information from Tobolsk about the advance of the Shestakov-Pavlutski Expedition, the Siberian Governor was working rapidly to evaluate the complex situation of the leaders, to plan the reorganization of the expedition with them and its implementation, and to determine any sanctions necessary. But most of the so-called decrees dated September 6, September 16, and October 3, 1728 reached the expedition at Eniseisk, Ilimsk, Irkutsk and

Yakutsk too late. The decrees of the Governor could not respond to the ever-changing situation. Not without reason, Shestakov, and later Pavlutski, usually made independent decisions without waiting for approval or affirmation of their actions. Gvozdev's journey over Siberian land and rivers took seven months. On June 29, 1728 the fleet arrived at Yakutsk. Ten years later in his testimony to the Siberian Regional Office, the geodesist recalled the dates inaccurately: "In 1728 I went from Tobolsk with navigator Ivan Fedorov and arrived in Yakutsk in 1729."[7]

Gvozdev stayed not more than a year in Yakutsk, the center of the expedition's accommodations and final provisioning. Here a month and a half after their arrival, they prepared to send out more than 200 people on various routes. The first to leave Yakutsk was a detachment of 23 people who went to Udsk ostrog by land. On eight rafts they floated all the supplies to Yudoma Cross and then to the East Sea. Two sailors and 159 recruits and Cossacks were there. Shestakov left on October 17, 1728 to supervise the servitors of the detachment who were to go to Okhotsk in the summer on river boats.[8] Twelve Yakutsk servitors under V. Shipitsyn were sent to Tashiverski ostrog, to Kolymski winter huts and then on to the ostrog of Anadyr. The important geodetic jobs, the study of the condition of the mouth of the Lena River and the soundings for depths in the delta during the end of September, a critical time period, were conducted by 24 Cossacks with supervisors Kozyrevsk, sailor M. Turnaev and G. Beliaev in charge.[9]

Besides the dispatch of the detachments from Yakutsk in August and September 1728, Shestakov continued to form the expedition, paying particular attention to food supplies, tools and materials and providing the servitors with salary and equipment. It's true that he often mixed his functions as head of the expedition with those of Cossack leader. As a result, the Yakutsk authorities censured him. Apparently several researchers have incorrectly concluded that Shestakov's presence in Yakutsk was of little value (56, pp. 9-10; 23, p. 48). In reality, it was because of his energetic and purposeful efforts that the formation of the expedition in Yakutsk proceeded slowly but surely in spite of bureaucratic obstacles and indifference and the petty tyranny and impunity of representatives of the local authorities.

From the sidelines, jealously guarding against the neglect of the smallest details in the interest of the expedition was the elderly Yakutsk Commander I. Poluektov. Shestakov also became involved in various affairs, often going beyond his authority and competence. One conflict arose after another. They were especially frequent during the formation of servitor groups, payment of their salary, the study of the natives' complaints and satisfaction of the various demands of the expedition. While evaluating the conflicts, one should keep in mind that Shestakov undoubtedly overextended his personal prerogatives and responsibilities. Poluektov could not separate the functions of Shestakov as the leader of Yakutsk Cossacks from his role as leader of the State expedition. In any event, he mistakenly considered the subordination of the Yakutsk leader as indisputable. All this caused, to a certain degree, the protracted nature of preliminary preparations of the Shestakov-Pavlutski Expedition. Or, as stated by the Siberian Governor, "great obstacles were put in the way."

According to historian Sgibnev, Shestakov, in spring of 1729, went to Okhotsk with part of his detachment. He sent the others by water with L. Petrov. D. Pavlutski remained in Yakutsk for a short time in Yakutsk before setting out for the Anadyr ostrog. Gens, Fedorov, Gvozdev and Speshnev delayed their departure for Okhotsk. This information needs correction. Shestakov headed to Okhotsk from Yakutsk, not in spring, but on June 22, 1729.[10] In Tobolsk on June 13, 1738 Gvozdev wrote that, Pavlutski dispatched he and Fedorov from Yakutsk to Okhotsk on June 7, 1729 (25, p. 46). In his autobiography of 1758, the geodesist recorded: "Upon my arrival at Yakutsk, Captain Pavlutski directed me, in 1729, to go to Okhotsk for making descriptions and preparing land maps" (52, p. 156). Pavlutski himself, on three occasions in his report of November 26, 1730 recalled his departure from Yakutsk. At first he complained about the absence of exact information concerning the treasury and the tools obtained by Shestakov. Then the Captain emphasized that, in 1729, he and Shestakov had not agreed and he left for Okhotsk "without agreement."

The note of the commander of the Anadyr Expedition about the departure of Shestakov from Yakutsk for Okhotsk together with the navigator, the assistant navigator, the geodesist and the apprentice, the grenadiers, the soldiers and servitors is not consistent with the traditional account that Gvozdev departed with Pavlutski.[11] But still the July 11, 1729 orders of Pavlutski to Gens, Fedorov and Speshnev most likely provide the most exact information about the departure of the expedition's detachments. Shestakov, with the servitors, headed for Okhotsk on June 22. On July 20, the navigator and assistant navigator together with the sailors received directions to be ready to move to the East Sea to join Shestakov (69, p. 73). On July 3, Gens and Fedorov prepared a list of the necessary navigational and skipper's supplies and tools for a single ship. On July 22, they added information about the number of carts that they would require. At last, on August 5, 1729 Pavlutski ordered Shestakov to move "without delay" to Okhotsk or anywhere the road led.[12]

On the way, on August 12, before reaching Yudoma Cross, Shestakov's detachment met with the returning First Kamchatka Expedition (11, p. 84). In 1732, Bering recalled some details about this meeting. Gens' poor condition, Speshnev's illness, and Gvozdev's absence especially struck him. "As for the geodesist, I didn't see him there," wrote Bering. "I do not know if he was dispatched elsewhere." Bering knew of Shestakov's dispatch of sailors "in three groups, which were to Kolyma, to the Lena River mouth and to the Uda River. If they are alive or not, I do not know, and (the Admiralty) does not know either."[13] On August 27, Pavlutski sent 60 Cossacks with Warrant Officer V. Makarov on rafts to the Lena River and then to Anadyr. He followed them shortly. In this way, in the summer of 1729, the two major detachments of the Shestakov-Pavlutski Expedition departed from Yakutsk to their support points, the Okhotsk and Anadyr ostrogs.

By the fall of 1729 Okhotsk became the main base for deployment of operations of the expedition on the mainland, Kamchatka and the Okhotsk and Bering seas. Until provisions arrived from Yakutsk, Shestakov hoped to use surplus food of the First Kamchatka Expedition left in Okhotsk by Bering. The ship *Fortuna* was useful in transport between Okhotsk and Kamchatka. Shestakov planned to sail the *Eastern Gabriel* to Penzhin Bay. After building ostrogs on the Penzhin and Olutor rivers, he was going to regulate tax collecting and bring "peace" to the Koryaks. Then, he would go to Anadyr ostrog by land.

Most historians who have evaluated the expedition consider (from our point of view, incorrectly) that all Shestakov's plans ended up in a fiasco (23, p. 48; 56, p. 17). Despite the harsh land and sea conditions, the Russian sailors and servitors completed many tasks with distinction. While completing their instructions, the land and sea detachments of the expedition conducted important historic geographical explorations and descriptions. In the summer of 1730, the son of the Cossack leader, Vasili Shestakov, and servitor Andrei Shergin were in charge of the *Fortuna* on voyages from Bolsheretsk to the mouth of the Kamchatka River and to the first four Kurile Islands. They described these places. On June 16, 1730 on the *St. Gabriel*, his nephew, Ivan Shestakov, also departed from Bolsheretsk. From August through October the "surpluses" from Yakutsk began to arrive as well. The major objective of the Cossack leader was the creation of a fleet of ships at Okhotsk. In his possession were the ships of Vitus Bering, includeing the *Fortuna*, built under the leadership of guard marine P. Chaplin and launched at Okhotsk on June 8, 1727.

Captain-knight Commander V. Bering laid the keel of the two-masted ship *St. Arkhangel Gabriel*, on April 4, 1728 in Nizhnekamchatsk and built under his supervision by the boat builder F. Kozlov. Launching occurred on June 9, 1728. In correspondence the name usually used was the *St. Gabriel, Garviil* and even the colloquial *Gavrila* or *Gavril*. The ship was about 18.3 meters in length at the keel, 6.1 meters in the width and 5 meters in height from sea level at the captain's bridge. The ship's equipment included a hold for cargo, crew quarters, officers' cabins and a galley. Okhotsk saw the construction of two ships in 1729: a large ship, the *Eastern Gabriel*, and a somewhat smaller one the *Lion*.

The *St. Gabriel* played the key role in sea explorations. Used for obtaining descriptions of the west coast of the Lamskoe (Okhotsk) Sea and of all the rivers flowing into the sea in the area of

Okhotsk ostrog, the *St. Gabriel* also explored the Shantar Islands, sailed along the Kurile ridge from Bolsheretsk, to Nizhnekamchatsk and on to the Big Land (the American continent).

The detachment headed by Gvozdev conducted the exploration of the mouth of the Uda River, the Shantar Islands and the mouth of the Amur, where it completed geographical measurements and described the western part of Okhotsk Sea and created maps (23, pp. 48-49). Several years later Petrov and Skurikhin informed the authorities in St. Petersburg and Irkutsk about the useful details of this voyage.[14]

In the fall of 1729, Shestakov himself departed from Okhotsk along the western coast of the Okhotsk Sea on board the *Eastern Gabriel* with seafarer N. Treska and a crew of 93 people. Shestakov reached the Tauisk ostrog with great difficulty. There in early October he learned that Petrov had reached Okhotsk from Yakutsk. Supplies for the expedition, Gvozdev, Fedorov and Speshnev were with him. Shestakov invited the geodesist to come to Tauisk. He appointed Petrov quartermaster and on October 15, directed him to prepare the ship's equipment and to build ships and boats in the mouth of the Urak River in the spring of 1730.[15] Nevertheless, Petrov headed for the Tauisk ostrog without waiting for authorization. He didn't meet with Shestakov until October 25, 1729 when the latter appeared in Okhotsk. Gens had to "make a brief stop" near Yudoma Cross, because of his illness. He left part of his freight there.[16]

The Cossack leader was in charge of the march by land through Koryak territory, departing from the Tauisk ostrog on November 17, 1729. There were 107 people in his detachment, including 17 servitors, Cossack Osip Khmylev "for writing," 30 deer Lamuts, 30 Okhotsk Tungus on foot, 10 Yakuts and 19 Koryaks. Following the *Eastern Gabriel*, the *Lion* went to sea under the command of *piatidesiatnik* (leader of 50 men) Ivan Lebedev. Missing the *St. Gabriel* and Shestakov in Tauisk, Lebedev took on his ship the advance group of Shestakov's land detachment with its leader Shestakov instructed Cossack sergeant Ivan Ostafiev "to record in the book what types of rivers there were and what size, the distance from one river to the other, where there were sufficient amounts of fish and other food supplies and what types of wild animals there were." He was also to "observe all things, pearls in shells, ores, stones, the color of the bones of mammoths and whales and to send samples with the reports if any of these were found."[17]

Shestakov's detachment joined Ostafiev's vanguard group and spent about six weeks at the mouth of the Gizhiga River waiting for the arrival of the Cossacks from Anadyr. Shestakov departed on March 9, 1730 without reinforcements. He crossed the Paren River and reached the Vakhla River. Then the detachment stopped and camped until March 13. On March 14, 1730 on the Egache River, between the Paren and Penzhin Rivers, Shestakov was killed.

Evaluating Shestakov's assignments, plans and actions from the very beginning of the organizing of the expedition, one should remember V. Berkh's fair estimate. He said that Shestakov was a clever and enterprising man and he could be useful in such a remote region. Müller called him "an illiterate charlatan" but added, "the enterprises of an uneducated man are often wiser than those of an educated one" (12, p. 7).

On March 11, from his camp in the Tylka River region, Shestakov sent his last direction on behalf of the expedition to Tauisk. He stressed the necessity for the crew of the *St. Gabriel* to travel to Kamchatka to the Bolshaia River, then to the mouth of the Anadyr, and then to the Big Land. Specifically it emphasized "if Gvozdev arrives, take him on the ship and grant him every consideration" (56, p. 15). Academician Müller, in his brief comments on Shestakov's authorization of March 11, 1730 regretted the absence of any information about the fulfillment of the Cossack leader's directions. But as far as Gvozdev was concerned, he noted, making a shift in the chronology of events, in 1730 the geodesist was already not far from Chukotka between 65 and 66 degrees north latitude. He was "on the coast of a foreign land, opposite dwellings of the Chukchi and found there people there with whom he could not converse for lack of a translator" (43, p. 404). This first publication about the sea voyage of 1732 with the mention Gvozdev's name appeared in print almost simultaneously with the end of the geodesist's career. Müller's inaccuracy in describing the events 25 years earlier may be explained only by the fact that Müller, like others in the Academy of Sciences, was not aware of Shestakov-Pavlutski Expedition documents in the Senate and Admiralty Board. This chronological error was frequently repeated in Russian and in foreign publications during the second half of the 18th century.

Before departure for Koryak country, Shestakov left a detail of servitors in Tauisk "for shooting game, building boats and repairing the ship" under the command of Cossack Trifon Krupyshev and grenadier S. Selivanov. When Petrov arrived in Tauisk, he also assumed command. Accordingly, in the winter of 1729-1730 the Tauisk detachment had a multiple governing authorities: shop assistant Tarabykin, Petrov, Krupyshev, Selivanov and, in short, anarchy. Shestakov's death (news of which was received in April 1730) further complicated the situation. Petrov, who illegally proclaimed himself chief commander, being afraid of the reaction to his bull-headedness and other abuses, tried to escape to the Koryaks and then tried to persuade the Cossacks to go to Kamchatka. Gvozdev, demonstrating courage and common sense, "arose against them and admonished the Cossacks to live in harmony and to go to Okhotsk, but the rebels kicked Gvozdev out of the meeting" (56, p. 22).

The report from Tauisk with the details about Shestakov's foray and his death was sent to Okhotsk for the Yakutsk regional office. It arrived in Yakutsk on May 24, 1730 simultaneously with a report from Gens and Fedorov, which had been brought by Speshnev. The sad news and the absence of any communication from Pavlutski prompted the navigator and assistant navigator to request the appointment of a new leader for the entire expedition. In the report it was stated: "Now in a crew without a commander, there is no leadership in self-appointed commanders, but only insanity and delays in all the meetings and quarrels." Each person wants the most important part for himself and we are concerned about the possible consequences.[18] Gens and his assistant were asking for responsibility of all the sailors "as was customary," and for a commander from Yakutsk so that everyone would be "under one command." He also asked for the discharge from leadership of all those who were appointed without authority, particularly Ivan Shestakov, "so that he couldn't take the death of his uncle out on anyone and also that the other commanders couldn't do anything to him."[19]

However, as soon as news about Shestakov's death reached Pavlutski, who was slowly moving on the way to Anadyr, he immediately ordered the concentration of the major forces in Anadyr. Gens, Fedorov, Gvozdev and Speshnev were instructed to sail immediately to the mouth of Anadyr with a crew. Pavlutski's instructions brought a certain clarity of purpose and subsequent actions were more goal oriented.

Navigator Gens in Okhotsk, upon learning of Shestakov's death, immediately sent authorization to Tauisk for Tarabykin and Gvozdev to "bring the *Eastern Gabriel* to Okhotsk with the servitors, because there is no other ship in which to sail to Kamchatka."[20] Okhotsk shop assistant Speshnev sent similar messages to Tarabykin and Petrov. He directed that the *Eastern Gabriel* not sail to Kamchatka first, but to Okhotsk for transporting government property to the Peninsula with equipment and provisions from Yakutsk arriving with V. Shestakov. Gvozdev was appointed commander of this voyage with seafarer N. Treska. In Tui, Petrov boarded the ship and read, in the presence of Gvozdev and Treska and the Cossacks the old authorization of Shestakov about the voyage to Kamchatka. Gvozdev, who undoubtedly had the respect of the servitors and whose opinion was respected, managed to convince everybody of the necessity of the voyage to Okhotsk and that it would be impossible to sail to Kamchatka and further to the Big Land without the navigator, assistant navigator, and the ship's craftsman, who were waiting for the Tauisk crew in Okhotsk. At last, on July 1, the *Eastern Gabriel* left Tauisk and on July 6, entered the mouth of the Okhota River.

Later, Gens in a detailed report informed Pavlutski that Gvozdev arrived from Tauisk in July on board the *Eastern Gabriel*, whose commander turned out to be Petrov: "And myself, I went to the skipper, boarding the ship you provided to me in your authorization from Yakutsk. I read it out loud to sailor Petrov and the servitors and asked sailor Petrov, if he was willing to obey me according to your authorization. And he, Petrov, refused to obey me together with the servitors, claiming that he had an authorization from Shestakov. And because of this we stayed in Okhotsk for a long time, waiting for further instructions."[21] On September 3, 1730 Gens took the list of all Shestakov's remaining servitors to Shestakov's commanders L. Petrov and S. Selivanov. On September 5, Petrov reported an inspection of all "food supplies received." The same day I. Shestakov arrived from Bolsheretsk and began the inspection of the *St. Gabriel*. This inspection lasted about a week. Some other curiosities were preserved in Gens' report to Pavlutski about the activities of 1730.[22] Among them, for example, is the thorough description of the characteristics of the *St. Gabriel*, the complaints he had about Gvozdev, Speshnev and other individuals with whom he had poor relations.

So, on September 4, Gens noted that the geodesist and the apprentice, in his absence, "made an examination of the servitors and taking revenge, placed the servitors under arrest for no reason." And "observing their attacks on the servitors, he refused to grant them authority" so that "nothing evil would happen." On September 11, during the inspection of the ship, Gvozdev and Speshnev said to him: "Give us a place on the ship, and if you do not, we shall move from Okhotsk."[23] Speshnev and Gvozdev based their activities in Okhotsk on instructions received from Shestakov and did not once obey the directions of the navigator. The navigator was particularly displeased with the refusal of the geodesist and the apprentice to provide him with servitors to work on the repair of the *Eastern Gabriel*. On this matter in a diary report, which turned into a kind of a conduct book, appeared new notes: "The geodesist Gvozdev refused and wouldn't let the servitors go to work. He said that he needed them to clean his own tools and that he lived in Okhotsk for the summer. But he did not prepare the tools. When the time came to do the work, there were even more obstacles. Apprentice Speshnev would not provide men for either work or guard duty until now. The mention of servitors not doing any work or guard duty because of Speshnev and Gvozdev tends to discredit their professions."[24]

On September 18, the *St. Gabriel*, under Gens' command, and the *Eastern Gabriel*, under Federov's, were ready for departure. On this day Gens had two entries: At the beginning, Cossack Ivan Tuev, after receiving the salary for the entire year of 1730, did not want "to serve on the crew, created great troubles for Gens and Fedorov in words and deeds, in order to stop the whole crew." According to Gens' explanation, this was not the first time that Tuev gathered the servitors together and denied the right of the navigator to be in charge of the ship, insisting that the navigator, the apprentice and the geodesist should be in charge together.[25] The second note concerned Speshnev and Gvozdev who, being on board the *Gabriel*, cursed the navigator and "humiliated and almost killed" him.[26]

On September 19, after Tuev was sent to the Yakutsk Office in chains and under security guard, both ships went to sea. That same day, Speshnev and Gvozdev declared to Gens that he should not be in charge of the ship. And "looking at them," complained the navigator, "servitors Iliia Skurikhin and Vasilii Yatskov asked for the reason that he, Skurikhin, had been punished by whipping. Also, the punishment of Yatskov by Speshnev and Gvozdev was not permitted." On the next day the following appeared in the record: "while at sea, the aforesaid Speshnev and Gvozdev called me, the navigator, a thief and other bad names."[27]

For the Okhotsk period, Gvozdev provides a very brief statement. He remarked about himself in his autobiography of 1758 that in "1730 by the authority of Captain Pavlutski, I stayed with the crew of navigator Gens." In Tobolsk, in 1738, the geodesist wrote: "We were in Okhotsk through 1730. In 1730 under the authority of Pavlutski, we sailed to Kamchatka via the Lamskoe Sea on the ship *Gabriel* with navigator Yakov Gens to Bolsheretsk ostrog..."

By that time the Anadyr detachment of the Shestakov-Pavlutski Expedition, which had departed from Yakutsk, passed the forbidding and "empty" regions, to "the Anadyr high land." Pavlutski and his detachment traversed the Verkhoiianski and Tai-Khaiakhtakh mountain ridges, the Alazeiski mountains, the Khalerchinska tundra, the Lena, Aldan, Yan, Indigirka and Alazeia rivers and many smaller ones. In April 1730 they reached Nizhnekamchatsk. He immediately tried to hasten the concentration of members of the expedition in the Anadyr region and to the mouth of the Anadyr River "for the purpose of exploration of the islands and the Big Land." Without doubt the head of the expedition and his seafarers had corrected the previous geographic concept of the Big Land as being opposite the mouth of the Kolyma (the map of Shestakov 1724). Those cartographic sketches that located the Big Land to the east of the Chukotka Peninsula guided them. They also had fresh information, obtained from the First Kamchatka Expedition of V. Bering and from Anadyr (30, pp. 142-152), where they arrived on September 3, 1730.

Despite the April 26, 1730 instructions to all those who had departed with Shestakov to the "Okhotsk road" (the navigator, the assistant navigator, the geodesist, the apprentice, the grenadiers, the soldiers and the servitors) to proceed to Anadyr, "nobody was coming" according to the report of Pavlutski to Tobolsk seven months later.[28] On October 9, the Captain sent new orders to Kamchatka, specifically ordering Ivan Shestakov to immediately take on the ship, Gens, Fedorov, Gvozdev, Speshnev, the sailors, and the servitors and to proceed the mouth of the Anadyr River.[29]

The first expedition of Pavlutski, March 15 to October 21, 1731 made significant contributions to the study of the remote Chukotka Peninsula, particularly to its hydrography and geography (14, pp. 117-118; 10, pp. 175-176; 34, pp. 158-160).

The images of the Russian people in the 18th century about Northeast Siberia in general and notably about Chukotka are reflected in documents of Yakov Ivanovich A. Lindenau, explorer of the region (3, pp. 286-310; 6, pp. 316-317; 21, pp. 244-247). The description (79, pp. 238-244; 4, pp. 156-157) and the working drawing (15, pp. 52-54; 47, pp. 129) came to the Senate on January 15, 1743 from Irkutsk. In his report of September 15, 1742 Irkutsk Vice Governor Lorentz Lang informed St. Petersburg about the sending of the working drawings with the description of the region, which he obtained on August 2, 1742 from Okhotsk Commander Anton Devier.[30] The work of Lindenau on the cartography of the Chukotka Peninsula was continued by Timophei Perevalov, a gifted person.

After the Chukotka Expedition of 1731, Pavlutski lead the expedition to the Perenski Koryakski ostrog on the Gizhiga River in March of 1732. From 1734 to 1739 he stayed in Kamchatka and from 1740 to 1742 he served as the Yakutsk Army Commander. Between 1744 and 1747 he organized several expeditions, which provided new and significant geographic information about northeast Siberia and adjoining regions.

The ships of the Pavlutski Expedition reached the coast of Kamchatka on September 30, 1730. Gvozdev and the members of the crew of the ship *St. Gabriel* disembarked at the mouth of the Bolshaia River. In 1738 he reported, "Together with the navigator and the assistant navigator we waited through July 1731 for instructions from Captain Pavlutski." From there "we received authorization to sail to the mouth of the Kamchatka River." The *Eastern Gabriel* approached the mouth of the Kykchig River but a storm carried the ship out to sea, damaging her severely. On October 4, the damaged ship was abandoned on the coast 30 versts north of the mouth of the Bolshaia River near the mouth of the Utka River. Fedorov, who was taken aback, sent one message after another: "There is no food, the servitors are asking to go back to town...many bears are approaching the ship and supplies...I don't even know how to get to town" etc, etc. Fedorov couldn't handle his subordinates in this difficult situation. On October 19, he reported to Gens about two people, who "left without permission and the rest of the servitors...are disobedient in work."[31]

The crew of the Pavlutski Expedition remained in Bolsheretsk "until summer." The most vital problems in the winter of 1730-1731 turned out to be the food supply and preparations for the voyage to unite with the Anadyr detachment.[32] Gvozdev and Fedorov were in charge of organizing fishing and hunting. "For hunting," additional gunpowder and lead were issued to the crew. Almost the entire crew participated in laying-in "fish food," since in addition to fishing, it was necessary to make fishing nets, obtain salt, and prepare the catch by cutting, drying, salting, smoking and jerking.

The lives of the crew in the ostrog slowly became normal. Gvozdev and Speshnev received quarters in the house of Bolsheretsk shop assistant I. Gerasimov. Under Speshnev's supervision, repair work began on the *St. Gabriel*. The apprentice himself conducted several reconnaissance trips in search of forests with lumber potential.

In these difficult conditions, Gens tried to solve all problems himself, considering himself to have the complete authority of the Cossack leader. He demanded from Petrov "seven sealed bags," addressed to Shestakov.[33] There was no agreement among the leaders of the expedition. It was not by accident that Speshnev said to Gens on behalf of himself and the geodesist. "If only you, the navigator, lived with us in agreement, then Bolsheretsk would be better, and not like it is now."[34] Assistant navigator Fedorov became seriously ill. On March 12, 1731 he reported, "My right foot has become lame. Not only can I not walk, but I cannot even get up."[35]

Thanks to finding the navigator's diary of 1730, like the report to Pavlutski, which consisted of 89 initial points, even the most unimportant events were scrupulously noted and preserved among the large variety of previously unknown details. The report Registered the day-by-day facts of the conduct of Gvozdev and Speshnev, painted in the darkest of colors. He was asking Pavlutski for "defense" against the two friends and in the end provided the most negative descriptions of each: "The apprentice Speshnev lives in leisure and doesn't assist in the building of the ship. The geodesist Gvozdev whether from Yakutsk to Lama or from Lama to the mouth of the Bolsheretsk River, made no calculations of versts. The creation of the land map was not accomplished until now in spite of the order."[36] As we shall see further, the accusations were not true: In most crucial situations both Gvozdev and Speshnev, ignoring danger, were in the center of the events, and conducted themselves with courage and dignity, while Gens himself, with various excuses, avoided the hard and risky tasks.

On December 28, 1730 soldier Klim Poleshaev brought the October orders of Pavlutski from Anadyr to the crew. Not having word on the state of the Kamchatka crew for a long time, the leader of the expedition emphasized to his subordinates the importance of their arrival at the mouth of the Anadyr River in time for further "exploration of the islands."[37]

The long expected orders hastened the preparation for the departure from Bolsheretsk. In the beginning of 1731, there were 112 participants of the Pavlutski Expedition in the Kamchatka ostrog. In addition, 26 members of Shestakov's crew were on the ship *Fortuna*, which according to Gens' opinion, "was not reliable for - treasury and people, for the ship was made out of twigs." Nevertheless, in April, N. Treska, the seafarer, was sent from Bolsheretsk to repair the *Fortuna*, which had been abandoned on a shoal near the mouth of the Kamchatka River.[38] As far as the *St. Gabriel* was concerned, the navigator thought the ship was quite fit for the Anadyr voyage. However, it was not possible to start with all the equipment and people "unless another ship were built." But they didn't build a new ship in Bolsheretsk, obeying the order of Pavlutski, "There will be no ship building in Kamchatka."

Gens suggested to Speshnev, who by that time had moved to Nizhnekamchatsk, that he transfer his crew of 15 people to F. Paranchin, who had been dispatched for preparing pitch and laying in "fish supplies."[39] In the meantime, the situation in the ostrog, with its laying in of provisions for the voyage, remained extremely tense. Thus, in January, Gens had not received the dry fish Yukola for the servitors. And now, the navigator continued, "There are not enough provisions which should be kept for the voyage and there is no fish supply for the crew."[40] Finally, L. Smetanin brought the salt, cereal and flour from the cache that had been accepted from I. Shestakov in 1730 and preserved on a ship in the harbor. On March 10, 78 puds 30 pounds of rye flour were issued "for provision of salary." Smetanin reported that by March 22 only two puds, 26 pounds of flour remained and for April, "I do not have any flour."[41]

In his next demand to the Bolsheretsk shop assistant on behalf of the supply of the ships, workers and nets, Gens solemnly concluded: "Already the springtime is coming and the crew is getting ready to move."[42] Nevertheless, regardless of all efforts, they managed to salt only one barrel of fish by spring. On June 12, Paranchin and L. Paliakov were sent with immediate instructions to expand the area of the search for fish and with this objective to organize trips of servitors to Verkhnekamchatsk and to the mouth of the Kamchatka River.[43] In the middle of May Smetanin and Petrov completed the repair work on the ship in the harbor. On June 8, the sailors were told that *St. Gabriel* was completely ready.[44]

On June 23, 1731 the *St. Gabriel* sailed from Bolsheretsk, around Cape Lopatka, and on July 9, reached the mouth of the Kamchatka River. The plan was to load provisions and sail without delay to the mouth of the Anadyr River. From there, part of the crew was to head towards the ostrog and the rest were to go to the Big Land. In three days Nizhnekamchatsk shop assistant, I. Kriskov, received an urgent demand from the expedition to provide 2,000 Yukola, 1,000 pieces of salted red fish or Siberian salmon and five interpreters for the voyage. But the reply was not consoling: In the state treasury "there were not any fish or provisions." According to local residents, in 1731 there were not enough fish coming from the sea into the Kamchatka and other rivers. That is why the servitors and local residents were having such "great difficulties and eating small fish." There was nothing to give to the crew. Krikov promised only to transfer the remaining rye flour from Bering's Expedition (69, pp. 75-76).

In the autobiographic notes of 1738, M.S. Gvozdev recalled how they "sailed the *Gabriel* by sea to the mouth of the Kamchatka River and were there 11 days." It was planned to finish all preparations for the voyage by July 20. The participants were preparing fish provisions, so that they could eat while at sea and were occupied with the maintenance of the ship (34, p. 57). The repair work was conducted under Speshev's supervision.[45]

As planned, on July 20, 1731 the crew boarded the ship and the *St. Gabriel* left for sea. Nevertheless, the ship stopped almost adjacent to the mouth of the Kamchatka River because of contrary winds and anchored. The voyage had to be cancelled. The opportunity to make the voyage was missed and the expedition crew remained in Kamchatka.

On July 20 the Itelman Rebellion began under the leadership of their chief Fedor Kharchin. It is described in some detail in the literature (36, pp. 487-500, 761-767; 34, pp. 47-81; 17, p. 45). Some additional information is presented from previously unknown writings by Speshnev and Gvozdev[46] of October 18, 1731. One should note that, in general, the admission of minorities of Northeast Asia into the Russian multinational state and the joining of Eastern Siberia to Russia was not always smooth. In its remote regions on the Okhotsk coast of the East Sea and in Kamchatka, tax collection, tyranny and unpunished Siberian administrators, and brutal methods of exploration sometimes took place in its most vicious forms. The new state citizens, *yasashnyeinozemtsy*, turned out to be the way of governmental colonization. This was accompanied by a number of negative occurrences, such as strict tax collection, severe and uncontrolled conduct of tax collectors, conscription, forced native participation in military marches and cheating during trade exchanges. Aggressive tax collection without proper regulation caused resistance among the local population. Their protests were expressed in their petitions, tax evasion and migration to remote, unreachable regions.

Until the end of July, the crew of the ship lived in town and then returned to the sea where the ship was anchored. A chapel and some residential dwellings were built on the island in the mouth of the Kamchatka River. From the dock, reconnaissance explorations of the river systems of Kamchatka, the Tigil, Ozernaia, Kharuzena, Uka, Avacha and others were regularly conducted in 1731 and 1732. The fall and winter of 1731-1732 turned out to be extremely difficult for the expedition because of malnutrition throughout the Kamchatka region as a result of the poor fishing season.

In October 1731, the German mining engineer, Simon Gardebol, who had signed a long contract with the Board in St. Petersburg[47] met the expedition.

Footnotes for Chapter 2

1. TsGADA, f. 248, Bk 666, p. 32.
2. Ibid., Bk 731, pp. 771-772.
3. Ibid., P. 1339, Bk 690, p. 340.
4. TsGAVMF, f. 216, Case 5, p. 96.
5. TsGADA, f. 248, Bk 781, pp. 707-708
6. Ibid., Bk 690, p.s 330-341; Bk 731, pp. 1335-1341.
7. TsGAVMF, f. 216, Case 24, p. 587.
8. TsGADA, f. 248, Bk 731, p. 1303.
9. Goldenberg, L.A., *The Pioneer Ivan Kozyrevsk in the mouth of the Lena.* Poliarna Zvezda, 1983, No. 4, pp. 119-126.
10. TsGADA f, 248, Bk 731, p. 1297.
11. Ibid., Bk 666, pp. 28-32.
12. TsGAVMF, f. 216, Case 3, P. 10; Case 4, p. 7, 10-11, 50, 109 110; Case 5, p. 32.
13. TsGADA, f. 248, Bk 664, p. 102.
14. TsGAVMF, f. 216, Case 24, pp. 485-487, 596-598; TsGADA, f.248, Bk 1102, pp. 1435-1438.
15. Ibid., Case 4, p. 30 and the other side of p. 30.
16. Ibid., P.s 8-9, Case 5, p. 32, No. 526; Case 24, p. 600.
17. TsGADA, f. 248, Bk 666, the other side of p. 32, p. 33 and the other side of p. 39.
18. TsGAVMF, f. 216, Case 4, pp. 31, 103-104.
19. Ibid.
20. Ibid., Case 5, p. 33.
21. Ibid., Describing "the troubles," A. Sgibnev noted that the expedition lived on the Okhotsk spit without anything to do. After becoming close friends, Speshnev, Selivanov and Gvozdev didn't listen to the shop assistant and scared him by breaking the Okhotsk customs. The navigator Gens became involved in the production of wine and selling it to the natives. Because of this, he moved 30 versts away from shore. He came back at the beginning of September with the arrival of the *St. Gabriel* from Kamchatka.
22. Ibid., pp. 33-34, 137-140 and the other side of p. 140.
23. Ibid., pp. 43-44.

24 Ibid., p. 39.
25 Ibid., Case 4, the other side of p. 118.
26 Ibid., Cases 181-182.
27 Ibid., Case 5, the other side of p. 44.
28 TsGAVMF, f. 216, Case 3, p. 22.
29 Ibid.
30 TsGADA, f. 248, description 113, Bk 1552, P. 1; Bk 1558, p. 151.
31 TsGAVMF, f. 216, Case 3, P. 38 and the other side, 155-156; Case 4, p. 72 and the other side.
32 Ibid., Case 3, pp. 81-82, 131; Case 4, No. 237; Case 5, pp. 65, 70-71, 86-88, 149, 173.
33 TsGADA, f. 1002, Case 1, p. 6.
34 TsGAVMF, f. 216, Case 5, the other side of p. 47.
35 Ibid., Case 3, p. 89.
36 Ibid., Case 5, the other side of p. 51.
37 Ibid., Case 3, p. 64.
38 Ibid., Case 4, p. 19; Case 5, other sides of pp. 58, 114, 188.
39 Ibid., Case 5, P. 51 and the other side of p. 91.
40 Ibid.
41 Ibid., the other side of p. 51, pp. 112, 183; Case 4, p. 71.
42 Ibid., Case 5, P. 88.
43 Ibid., p. 190.
44 Ibid.
45 Ibid., p. 90.
46 TsGADA, file 248, Bk 666, pp. 298-303.
47 Ibid., Bk 180, the other side of pp. 72-75, pp. 204-207; Bk 664, pp. 88-89; Bk 690, pp. 313-328; Bk 773, pp. 150-151.

Chapter 3
At the Head of a Sea Expedition to the Big Land in 1732

The spring of 1732 arrived. On February 11, before traveling from Anadyr to the Gizhiga River, Pavlutski sent new instructions to the leaders of the Kamchatka detachment. Referring to this directive, Gvozdev wrote, "We received instructions to sail with the navigator and assistant navigator on the ship *Gabriel* around Cape Kamchatka to the mouth of the Anadyr River and then to the Big Land opposite Cape Anadyr, checking the islands, counting them, and determining if people lived on them" (53, p. 90).

On July 20, 1731 several hours before the scheduled departure of the *St. Gabriel*, Fedorov informed Gens that for the past four months he had completely lost the use of his right foot. He said, "On a voyage with such pain from the sea air, it is very difficult and I cannot do anything." The assistant navigator asked not to be sent on the Anadyr voyage but to be left in Nizhnekamchatsk and that the tax office furnish him with housing. He also asked that his regrets be conveyed to Speshnev, Gvozdev and the other seafarers. Fedorov's letters that followed on August 8 and 11, September 29 and October 2 and 14 are full of references to imaginary and actual mistreatment and requests for help.[1]

When cold weather came, the assistant navigator found himself in a difficult predicament. All apartments in the town were occupied. Abandoned by everyone, lonely, without food and supplies of any kind for the winter, Fedorov had to remain in a cabin on the ship. He still couldn't use his right foot. He repeatedly and pathetically appealed to Gens for sympathy: "Now you are wearing me out by the winter cold and starvation on board the *Gabriel*. I am suffering very much. It is difficult to be on the ship without food and heat in my condition."[2]

By a written order of February 11, 1732 Pavlutski relieved Gens of his position as leader of the detachment: "You, the navigator, living in Kamchatka, are interfering in affairs outside your authority while you ignore the matters with which you have been entrusted. You live in Kamchatka in leisure, filled with your caprice and not being of any use to the detachment." He ordered that Gens transfer all servitors, soldiers and supplies to Gvozdev. As for the navigator and assistant navigator, they were only to be in charge of the sailors and were told, "not to participate in any other activities because it is known that you are quite poor sighted, your feet are hurting and you are unable to perform the duties entrusted to you." The Captain's directive also required the navigator and assistant navigator, in the event they recovered from their disabilities, to sail on board the *Gabriel* with the sailors and geodesist Gvozdev but otherwise to remain in Nizhnekamchatsk until they recovered.[3]

On May 1, Gvozdev received the same type of directive as the one dated February 11. On that date Gvozdev informed Gens that Pavlutski "commanded that Gvozdev receive the soldiers and a certain amount of guns, gunpowder, lead and a variety of food supplies and that Gvozdev was in charge of the soldiers and servitors. The navigator and assistant navigator were in charge of the sailors only and had no authority over anything else." Gvozdev also emphasized Pavlutski's additional orders to take on board the *Gabriel* the seafarers who were at sea with Bering and those sailors who were with Gens and Fedorov in case the two left Nizhnekamchatsk. In addition, Gvozdev emphasized that Gens was to give the taxes collected by the crew to the Commissioner.

As the head of the voyage to the Big Land, Gvozdev ordered Gens to direct the servitors, seafarers, soldiers and sailors to transfer all the food supply, surpluses, oil, the *St. Gabriel*, guns, gunpowder and lead as soon as possible.[4] In his report of May 3, 1732 to Pavlutski, Gens said he transferred the ship, all the supplies, the soldiers and the servitors. He complained that Gvozdev also took the sailors, without authorization, leaving them without a single person under their authority.[5]

Also preserved is a note from Gens to the geodesist regarding the completed transfer of the ship, crew, weapons, food and accounting of the treasury. "Because of our illness we will be left behind in Nizhnekamchatsk," wrote the navigator, "so take the sailors on board the *Gabriel*." Gvozdev's placed his counter signature on this document: "received by geodesist Gvozdev."[6] This document provides us with the general description of the ship, the condition of its rigging and food supplies and other property transferred to Gvozdev. According to the personnel count of June 11, 1732 Gvozdev's command consisted of 145 men: 125 servitors, 4 seafarers, 7 sailors, 4 grenadiers and 5 soldiers.

Of special interest are eleven directives with Gvozdev's signature issued to Petrov, the quartermaster.[7]

On the second day of his appointment, May 2, Gvozdev gave his first directive. Petrov was to issue Speshnev two destei of writing paper for keeping notes concerning the treasury. The orders of May 5 and June 2 were given for the purpose of conducting an inventory of property and supplies.

On May 31, Gvozdev directed A. Shering be issued seven pounds of gunpowder "for protection while collecting tax." On June 13, Sheshnev received iron and tools for building a boat. On July 6, he ordered 10 cannons and 50 cannon balls issued for security and nets used in gathering wood.

In his last letter to Gens before departing to sea, Gvozdev said, "On July 8, I will be going to sea with the servitors. By the time you, the navigator, receive this letter you should be ready for the appointed date for the voyage. If you haven't recovered from your illness, then you should respond in writing immediately."[8] For reasons unknown, the departure did not take place as planned. On July 13, Gvozdev provided servitors E. Permiakov, I. Portniagin and L. Poliakov with several pounds of gunpowder from the *St. Gabriel*.

With the removal of the navigator and assistant navigator from participation because of illness, Gvozdev became the leader of the voyage to the Big Land. At least he had carried out all the organization and preparation necessary for sailing. This documentation has an important bearing on the controversy. It is known that Gens remained on shore, but Sgibnev insisted that Fedorov be, in accordance with the order of July 13, brought on board "against his will" and died soon after his return. It is presumed that Fedorov's ailment made him irritable, nervous, weak, and often helpless. He was not capable of fulfilling his duties during the voyage.

Unlike the weak and insecure Fedorov, Gvozdev proved himself to be a resolute commander who kept his wits about him even

GVOZDEV'S LETTER OF MAY 1, 1732

in the most difficult situations. Gvozdev no doubt had strong character and the respect of the servitors, although he was firm and acted according to regulations. Of interest is the denunciation of the geodesist by Cossack F. Tibalov, submitted to Fedorov on June 1, 1732 for forwarding to Pavlutski. Tibalov related that Gvozdev, disagreeing with an accounting of funds, kept him under security on the ship, "ordered me to disrobe, tied me down and had me whipped almost to death."[9]

There is no doubt that Gvozdev was the head of the 1732 voyage. Without any documented information, some historians had based their presumption of Fedorov's leadership on emotions only. Most historians have accepted this mistaken conclusion without verification. It is obvious that the seriously ill assistant navigator did not play a very important role before, after or during the voyage itself. On shore and on board the ship, the responsibility lay on the shoulders of geodesist Gvozdev, commander of the *St. Gabriel*. His actual assistant in navigation was seafarer K. Moshkov. On April 8, 1741 I.F. Skurikhin, a participant in the voyage, confirmed to the Okhotsk port office that Gvozdev was commander of the ship (30, p. 236).

Not a single document of the Shestakov-Pavlutski Expedition and voyage mention Fedorov as commander of the *St. Gabriel* in 1732. In researching the voyage to America, all scholars encounter an absence of information about Fedorov. Nevertheless, only Sokolov took the trouble to highlight this important fact. On the relations between Gvozdev and Fedorov, all that exists are different points of view that are often contradictory.

First, A. Polonski stated (51, p. 395) that Gvozdev, after the transfer of the *Gabriel* from Gens who was left sick in Nizhnekamchatsk, sailed according to instructions. Fedorov joined the ship with a tremendous ulcer on his foot that caused his death in February 12, 1733 shortly after his return. A. Pokrovski (69, p. 13) ignored the important role of Moshkov on the voyage. It was the view of F. Golder that Gvozdev had full responsibility, assisted by Moshkov (75, p. 158). The unsubstantiated statement that Fedorov was the commander of the ship and Gvozdev was his assistant became very popular.

Though this conclusion differs with the historic facts, most historians accept it. Thus, Sokolov writes that because of Gens' illness and blindness, the assistant navigator was in charge of the ship. His assistant and friend was the geodesist, Gvozdev, who was "healthier and more active." Sokolov wrote that Fedorov "who suffered from a foot ailment was not prevented from fulfilling his service and seemed to dominate Gvozdev." This phrase is not written properly and its meaning is unclear (59, p. 81). According to Sokolov's statement, it turns out that the assistant navigator could do his duties. This was only during the voyage itself because beforehand and afterwards, he was always ill. Fedorov, who had been carried onto the ship, preferred living in a warm cabin to starving without housing or help in town.

"Fedorov was the commander and Gvozdev was the assistant," insists academician L.S. Berg. He agrees with M.I. Belov who added that the geodesist was fulfilling the duties of the navigator. These statements, provided without any documentary support, are especially tenuous because of the peculiar contradictions. In describing the sailing, Berg mistakenly stated that, on August 7, Fedorov visited the coast, while Belov claims that on August 7, Fedorov could not even get out of bed (9, p. 99; 6, p. 260). The statements of A.V. Efimov can also be confusing. He said since Gens became almost completely blind, he entrusted the expedition to Fedorov, who was ill with scurvy and suffered from wounds on his feet. He added that Gens ordered the sick Fedorov brought to the ship by force (30, p. 160; 27, p. 222). But as is well known, Gens was not in a position either to command or commission anything because of Pavlutski's order. On February 11, 1732 Gens was relieved of his duties as commander and replaced by Gvozdev.

By his explicit statements, it is known that V.A. Divin also recognized Fedorov as commander of the ship. Gvozdev, as his assistant, was supposedly in charge of "the scientific portion." Divin also considered both seafarers had equal rights, but noted that there was no agreement between them. He suggested that while Fedorov understood the sea very well, Gvozdev had only theoretical knowledge but had no practical experience. For this reason, Divin concluded that the navigator treated the geodesist with scorn, often preventing him from fulfilling the tasks which he himself couldn't accomplish because of poor health. However, the collected information doesn't allow such a conclusion. Other than the loss of the ship *Eastern Gabriel*, Fedorov's insecure actions on shore after the shipwreck, and his prolonged illness, there has not been any information found regarding his qualifications and knowledge as an assistant navigator. In his 1956 work, Divin considered Fedorov's merits in light of his successfully operating the ship in complex navigational conditions because of his good training and that of the sailors. Paradoxically, the same author, in another book published in 1971, attributes the same qualifications to Gvozdev (25, pp. 21, 26-27; 24, pp. 79, 83, 85).

The search for preconceived ideas about Gvozdev brings one to the opinion expressed by Sokolov who disliked the geodesist. For historians, the difficulty in evaluating the relations between the leaders of the voyage is complicated by the fact that Gvozdev's original reports are preserved while Fedorov's documents and other testimonies on his behalf were not. Accordingly, Sokolov called the report of the geodesist "muddle headed." He considered that Gvozdev's name became known only because of the documents that reached us. Those developed by Fedorov "were lost." He suggested that it might be that the assistant navigator would not verify Gvozdev's later reports. The historian was struck by Fedorov's proud words addressed to the geodesist: "He did his job and did not ask for any help at sea and will not ask for any in the future from the ranks of those who are not knowledgeable of the sea and the assistant navigator's profession." It seems that this statement was not made out of pride but rather out of the stubbornness and helplessness of a wounded, suspicious and bitter man.

The *St. Gabriel* began its journey with the geodesist Gvozdev, seafarer Moshkov, assistant navigator Fedorov, 4 sailors, 32 servitors and the recently baptized interpreter, Egor Buslaev.

According to Gvozdev's reports, the voyage began on July 23, 1732 when the *St. Gabriel* left the mouth of the Kamchatka River. After four days they rounded Cape Kamchatka and headed for the Chukotka Peninsula. According to Efimov, contrary to the documents presented, the expedition sailed out of Bolsheretsk and rounded "Cape Kamchatka" or Cape Lopatka on July 27 (30, p. 165; 27, p. 227). Gvozdev's 1743 report states: "We sailed the same year, 1732, on July 23 from the mouth of the Kamchatka River and went around Cape Kamchatka on July 27 and arrived at Cape Anadyr on August 3. We sailed from Cape Anadyr in search of islands."[10] The basis for the search for an island in Bering Strait was the testimony of Moshkov, who had sailed in the same area with Vitus Bering. As it is known, on August 17 (August 16 on the civilian calendar), 1728 Bering's First Kamchatka Expedition saw an island named in honor of St. Diomede (known in Russia today as Ratmanov).

The *St. Gabriel* sailed along the Asian coast to the north. On August 5, it arrived on the south side of Cape Chukotka, anchoring three versts from shore for lack of wind. Historians of the voyage disagree about which Cape was described. Some suggest it was Anadyr (the cape now called Chukotka), which is not Cape Dezhnev but one of the capes located north of Cape Chaplin (9, p. 99; 25, p. 21). Most of the research indicates that Gvozdev's expedition stopped somewhere south of Cape Dezhnev (23, p. 51; 24, p. 80). On Spanberg's map in 1743, the Asian bank of the Bering Strait is not marked Cape Anadyr (in the south part of the western coast) and Cape Chukotka (or eastern, Cape of Dezhnev), but is indicated as $66°08'$ north latitude. Therefore, we can assume that the first stop of Gvozdev's expedition was somewhere between Capes Dezhnev and Lütke.

On the same day, Gvozdev sailed with his crew to examine the coast. The travelers saw "an empty place," a small river, a herd of the deer and two natives. During their first reconnaissance tour, they increased their supply of drinking water (two barrels) and food provision (two deer). According to Gvozdev, there was no chance to communicate with the natives because they ran away and hid in the cliffs. On the following day two baidarkas approached but they did not come very close to the ship. Attempts to converse with them through the interpreter failed. The Chukchi men did not respond. After staring at the ship, they returned to shore (52, p. 151).

Terminus of Cape Mountain (Photo by J.L. Williams)

Razorback Mountain   (Photo by J.L. Williams)

Cape Mountain   (Photo by J.L. Williams)

Three Diomede Islands   (Photo by J.L. Williams)

On August 7, Gvozdev and the sailors went ashore again, continuing their observations. They examined abandoned dwellings of the Chukchi: two empty yurts that were built into the ground from old and torn apart whale bones (52, p. 152). After staying four days near the Chukotka Peninsula, a good wind began to blow. They raised anchor and the expedition continued the search for islands. Sailing by the route established by Moshkov "along the south side of Cape Chukotka to the Anadyr side," was not successful (52, p. 152). Without navigational guidance and relying completely on his own memory and intuition, the seafarer apparently made an error in his selection of general direction. In such conditions Fedorov told Gvozdev that in his opinion they were searching in the wrong area. Only one not aware that Gvozdev was the leader of the voyage and commander of the ship could share Efimov's confusion regarding this message (why the assistant navigator wrote to the geodesist). Gvozdev and Fedorov agreed to stop sailing south on August 9. They turned around and returned to the place of their first mooring, where they sent the sloop for more drinking water.

On August 11, under fair winds, the ship sailed into the sea. But after two days they returned to the Asian shore where they saw yurts on a river during the calm weather. According to sea traditions, Gvozdev was in charge of search and exploration of new places. This time again, the geodesist with sailor L. Petrov and the servitors, examined "the yurts that were built of wood into the ground" (52, p. 152).

In a short period of time Gvozdev observed the peculiarities of the household, way of life and conditions of nature on the Chukotka Peninsula. He noted, for example, that Chukchi "for their nutrition ate whales and walrus and didn't have any other food." He also was struck by the absence of trees - "only tundra" (52, p. 152).

So, from August 5th through the 15th, the crew of the *St. Gabriel* sailed along the shore, disembarking at several locations of the peninsula. They sailed for a certain distance to the south and to the east, observing the nature of the region, and trying to establish contact with the native population.

At 11 o'clock on the morning of August 15, with fair winds "the anchors lifted and all sails set," the seafarers "went on their way." Gvozdev didn't state the exact date of their arrival at the first island. He did indicate that on August 17, at 7 o'clock, he examined an island, most likely Ratmanov Island. But the wind changed and the boat began tacking, so they returned to the Asian mainland. During the calm, they lowered the sloop into the water. Gvozdev, Petrov and 10 servitors sailed to the yurts that they had seen on the shore. Returning to the ship, two baidarkas followed them and almost caught up them. In each baidarka there were twenty natives. Through the interpreter, Gvozdev tried to obtain information in support of the voyage's purpose. But they did not respond to his question, "What type of people live on the island?" The people in the baidarkas said only that they "were large toothed Chukchi" and "that their homes were on Cape Chukotka." Evaluating this information, Berg commented that the Eskimo people didn't call themselves "large-toothed."

This might have been their Chukotka name if by 1730 they had not stopped wearing the utulki. The Asian Eskimo called themselves "yuty" or "yupiyt" (which means "people" or "real people"). Berg concluded that Gvozdev was on Cape Dezhnev near the Eskimo settlement of Nuukan (9, p. 100). From there the *St. Gabriel* again sailed to the first island (Ratmanov Island), approaching it from the north. They noticed dwellings on the island. In the sloop, Gvozdev, Petrov, and 10 servitors approached the shore and met with the residents, who said they were Chukchi. Berg, who found this statement strange, remarked that the local Eskimo called themselves Chukchi. He explained that the interpreter, E. Buslaev, who was taken from Kamchatka, might be Koryak by origin. He could easily converse with Chukchi, but not with Eskimos. The Asian Eskimo and those from the islands discovered by Gvozdev understood the Chukchi language. It is possible that in his description, Gvozdev called all natives he saw Chukchi. Thus, during inquiries about the Big Land, the residents of the island didn't say how big it or the island was, but only that "our Chukchi lived there" (52, p. 153).

Going ashore and exploring two abandoned yurts ("in the ground made of wood from spruce trees") the seafarers discovered "food only from walrus and whale." The settlement on the south side of the island turned out to be more populated. There were about 20 yurts. The Russian detachment under Gvozdev in the sloop and the interpreter in a baidara approached the shore. All attempts to obtain information about the Big Land from the islanders were in vain. Of great importance, Gvozdev present the descriptions and details regarding their visit to the first island and the Big Land. He also reported that the island itself "was not large and that there were no trees on it" (52, p. 193). The report of 1741 presented the measurements of the island: the length was 2.5 versts and the width was 1 verst.

On August 20, the ship anchored near the second island, now called Kruzenshtern. I.F. Skurikhin, who observed both islands from the ship from 1.5 to 2 versts distance from shore, said that both islands were no larger than 2 or 3 versts. E. Buslaev was sent to the island in a baidarka and E. Permiakov went in the sloop with 10 Cossacks. They all returned quickly since there were no negotiations with the natives. According to Gvozdev's description, the second island was smaller than the first. He provided different information on the distance between the two islands: in 1741, 1.6 kilometers and in 1743, 926 meters. V.I. Grekov calls the last measurement closer to reality (23, p. 347).

In 1779, almost 50 years after the voyage of *St. Gabriel*, Cossack Centurion Ivan Kobelev visited both islands. He called them Imaglin and Igellin. According to his description, the distance from the Asian shore to the first island was 40 versts. The island was 2 versts wide and 5 versts long. There were 398 people in two settlements 3 versts apart. Igellin Island was 3 versts long and 1.5 versts wide with 164 residents. According to Kobelev, both Chukotka and America can be seen from the second island, which is about 30 versts from the North American coast. Major M. Tatarinov, the famous Siberian cartographer and a participant in the Nerchinsk Expedition of F.I. Soimonov, while preparing an extract from Kobelev's journal, claimed that the Cossack was the first Russian to see from the islands of Imaglin and Igellin, the eastern end of Asia and the shores of North America at the same time (30, p. 232). Thus by the end of the 1770's, the width of the narrowest part of the Bering Strait was considered to be 82.1 kilometers. We cannot agree with Tatarinov. Long before Kobelev was in the region of the Diomede Islands, Asia and America might have been seen by S. Dezhnev in 1648 and by the participants of Gvozdev's expedition in 1732.

It is possible that Gvozdev might not only have seen the Arctic and the East Seas, the wide strait and two islands, Ratmanov and Kruzenshtern, from the capes of Chukotka, but also heard the ancient legend of the natives about the origin of the Bering Strait.[11]

Besides his reports and stories, which were possibly exaggerated, Gvozdev himself was especially active in the expedition in the Bering Strait. I.F. Skurikhin, one member of the expedition, verified the Gvozdev's activities. On more than one occasion, Gvozdev sailed from the ship to the shore to explore and describe the different geographic regions of the Chukotka Peninsula and Ratmanov Island. He tried to obtain necessary information about the Big Land and the islands from residents of the islands. During this period of the voyage there is no mention of assistant navigator Fedorov. That is why the explorer of the Polar Seas, V. Yu. Vize, was in error when he claimed that the assistant navigator was the first Russian who had disembarked on the islands of Ratmanov and Kruzenshtern (16, p. 63).

Gvozdev's record of sailing the next day, August 21, 1732 should be written in gold letters in the history of geographic cartographic study of the East Sea. On this day the first European reached the northwest coast of North America. In the 1743 report he wrote: "On August 21, at 3 a.m. a wind arose. We lifted the anchor. The sails were set and we sailed to the Big Land. When we arrived, we stopped at anchor opposite a place where there were no dwellings" (53, p. 153). The same event was reported by the geodesist earlier and in 1741. The differences in the reports are as follows: the 1741 report speaks of the "helpful" wind and the location of the anchorage, four versts from land, and a different time of departure from the island to the Big Land, "at 3 p.m."

Spanberg's map of 1743 depicts the route of *St. Gabriel* from the Diomede Islands to the Big Land very generally to the east-southeast. The first stop near the shores of Alaska and all further changes in the route are not presented. In the meanwhile, very nicely depicted is the shoreline of the contemporary Seward Peninsula to the northeast, and to the southeast is another cape of the American continent, Cape Prince of Wales. To the northeast, the shoreline reaches to 66°45' north latitude, that may point us to the possible first anchorage of the expedition, since the further movement of the ship was mainly in a southerly direction. It so happened that while on duty during one of his few watches, the sick Fedorov, "without mutual agreement" as Gvozdev recorded, ordered the anchor raised and the boat sailed "near the land to its south end" (52, p. 153).

There is no doubt about the ship's approach to Cape of Prince of Wales, that is also proved by Spanberg's map. Then the seafarers "near the south end saw to the southwest inhabited yurts about a verst and a half apart. "Berg assumed this was the Eskimo settlement on the cape, called Kingegan on contemporary maps. Contrary winds did not allow the *Gabriel* to approach the shore. That is why they continued sailing "next to the land on the south. The depth of the water became shallow. When we reached a depth of 6 to 7 sazhens, we began to tack, turning back so that we would not get very far from land." This information about their location is the same as the 1741 report which stated that in the shallow place "the lead was dropped at 6 or 7 sazhens and we began to tack near the Big Land so to approach the land." A strong north wind prevented the *St. Gabriel* from approaching the land. It is precisely because of this that the movement along the American shore to the southeast from the Cape of Prince of Wales was accompanied by tacking the ship. The gradual reduction of the water depth to 6 or 7 sazhens (12.7 - 14.9 meters) and fewer is typical for the entrance into the Kaviaiak Bay. Spanberg's map marks the shore line of the Seward Peninsula as 65°15' north latitude, which is very close to the latitude of this bay. Most likely it is from here that the ship headed away.

With a heading of southwest and a strong north wind, Gvozdev's expedition approached a fourth island on August 22, "and because of bad weather, we couldn't drop the anchor near this island. We were carried from it with the sails down" (52, p. 153).

It has long been known that the fourth is land described by Gvozdev is King Island or Ukivok, which is located not far from the coast of Alaska. The *St. Gabriel* undoubtedly followed a westerly route, since departing to the southwest from Cape Spencer or Kaviaiak Bay it couldn't miss King Island.[12] This circumstance virtually proves the location of the first Russian ship near the shores of the North America. That is why one cannot assume that the ship advanced along the American coast to Nome and then returned to Kamchatka. The expedition sailed to Nizhnekamchatsk from the fourth island, which is located not to the southwest from the Nome Cape, but to the northwest. The assumption of some researchers that the *St. Gabriel* possibly advanced along the American shore to Norton Sound doesn't coincide with the information presented (81, p. 5).

A lively discussion has emerged among scholars about the absence of any mention of a third island in Gvozdev's reports. The geodesist writes about the first island, the second and suddenly moves to the fourth. Most researchers conclude that the Gvozdev expedition of 1732, while observing the Big Land, thought it was a third island (51, p. 392). F. Golder came to a similar conclusion. To prove an identical view, Efimov presented a lengthy argument. First, he emphasizes a quotation from the journal of the ship by Gvozdev and Fedorov, calling the Big Land, the big island in 1732: "On August 22, about midnight, from the island (fourth island L.G.), a man visited them who Gvozdev asked through Egor Buslaev the interpreter what type of people lived on the island. The man answered that Chukchi lived on the island. Gvozdev also asked if there were any beasts and forests on the big island. The Chukchi said: on this island there are the foxes and martens and forest" (51, p. 392; 9, p. 102; 30, pp. 168-170). Efimov considers this to be the only known piece of information from the diary that finally solves the question about the third island by characterizing the images of Gvozdev and Fedorov of the Big Land as the big island. But by this interpretation and trying to prove the point for his supposition, his analysis of the document leads to the truth of the matter. Thus, he states that in the presented extract "The Big Island" is written in capital letters as well as the Big Land. This comment, regretfully, is not true. Checking the original document helps us to establish that "the big land" is written twice with the small letters and most likely is not the synonym of "Big land." Besides this, in the extract from the journal, the big island is also called Cape Chukotka.

According to Efimov's explanation there was a place on the American shore, called the same as the Cape on the Asian shore. V.I. Grekov completely rejects the thesis that "The Big Land" is the third island. The report of I.F. Skurikhin in 1741 shows the participants of the expedition were familiar with different geographical areas and understood the difference between an island and a mainland. At the Okhotsk office, the Cossack said: "and half a verst before reaching it (the land - L.G.), we made out that it was not an island, but the big land, the shore of yellow sand. There were dwellings of yurts along the shore and a lot of people walking along the land. The woods on this land were great: Larch, Spruce, Poplar, and a great number of deer" (53, p. 101).

During the exploration of the north side of the first island (after August 17), Gvozdev waited for but didn't receive an answer from the islanders to his questions about the Big Land, "how big or how many islands." The reason King Island is called the fourth island is because V.I. Grekov found on Spanberg's map of the strait not only the first and the second Diomede Islands but also a third small, uninhabited island. Since there were not any events connected with it, the ship did not approach it and Gvozdev did not mention it in his report. In this way, according to contemporary geographic nomenclature the first three islands are the Islands of Ratmanov, Kruzenshtern and Fairway Rock. The fourth island is King Island (or Ukivok).

By the 1730's, information collected on Chukotka from the natives about the Big Land referred to it as an island. Even in 1760's when Anadyr Commander F. Plenisner was collecting information about Chukotka, the islands in the Strait and the Big Land, geographical perceptions of the Chukchi and the Anadyr Cossacks had not changed very much. Such is, for example, the record about the island opposite Cape Chukotka "on which the Chukchi are living, who are also called 'Big Teeth...' and beyond this island there is also an island which is called the Big Land and it is considered to be well populated, the people call them Kykhmyltsy."[13]

In this regard, the perceptions of the crew of the *St. Gabriel* are not original and Gvozdev and Fedorov followed the traditional views in 1732. However, 9 and 11 years after the voyage, while preparing the new reports, without doubt, the geodesist wrote about the Big Land as a continent and not as an island. At the beginning of 1740 during the designing of a new expedition, the Commander of the Okhotsk Port, A. Devier, proclaimed that the Big Land, most likely, is America. Still earlier, in 1736 the exiled Captain-Lieutenant V.I. Kazantsev reported to Tobolsk that "opposite Cape Chukotka, on the east side, is Cape America."[14]

When on August 22, the *St. Gabriel* drifted away from King Island, Sailor L. Smetanin, with the other members of the crew, addressed Gvozdev with a request to return to Kamchatka, since "it is getting late and the winds are becoming very powerful" (52, p. 154). At that time, an Eskimo in a small boat, a kukhta, approached the ship from the island. According to the description by Gvozdev, the kukhta is "all leather including the top, but seats only one person. He wore a shirt over his clothes. The shirt is made from whale gut and tied around his kukhta, hands and the head, so that the water couldn't penetrate. When the sea wave splashed him all over, the water can't get into his kukhta. Also, there is a large inflated bladder tied on his kukhta so that sea waves cannot turn it over" (52, p. 154). When asked about the Big Land, the islander said "our Chukchi" live there and the forest "is Spruce" and "there are deer, marten and fox and river beaver." Presenting this information on the flora and fauna of Alaska, Gvozdev also emphasized that on Cape Chukotka and on the first and second islands "there were no forests" (52, p. 154). The absence of forest was also typical of the fourth island, the general view and size of which the geodesist determined almost exactly. King Island is 2 miles (3.7 kilometers) in circumference. His description was "small and round." In the report of 1741 he wrote, "small and round in its width and its length, 4 versts."

I.F. Skurikhin's memoirs, that shift places, supplement Gvozdev's description. Thus, the exiled Cossack said, "one naked man swam to us on a bladder" not from the island, but from the Big land and "showed them where to look for Kamchatka and invited them to his land, promising to provide them with food." Skurikhin's story about the trade between the geodesist and the native, not presented in Gvozdev's report, is of great interest. Skurikhin said that Gvozdev threw a stick to the arriving Eskimo, containing a gift of needles, thimbles and little bells. The islander accepted the present, sailed to the shore, and then returned to the ship. At a distance of 30 sazhen, he left "a leather fish bladder" on the water and then sailed away from it some distance. He returned to shore only when he saw a sailor from the ship swimming to the bladder.

Inside the bladder brought to the ship "appeared the present: two marten parkas and a large quantity of fox skins." According to Skurikhin, "the parkas and foxes were divided between Gvozdev and the assistant navigator because 'the present' to the islander came from their private goods not from public goods" (53, p. 102). The location of an event in this second account also doesn't coincide with Gvozdev's description. Skurikhin claims that on the return to Kamchatka, the expedition stopped at one of two islands that "they had seen during sailing." Gvozdev and servitors I. Rebrov, I. Zalevin, G. Nekhoroshikh, M. Sharypov, D. Shchadrin, E. Permiakov, V. Zyrian and sailor L. Smetanin visited this island in the sloop. After four hours on the island they returned to the ship with "about 10 or 20 slaughtered deer, nedorosti and young reindeer," that they divided into 40 parts. Every member of the crew received "two or three pieces." They also brought about 100 puds of walrus tusks to the ship. On arrival in Kamchatka, "the tusks" were sold to an elderly man at the monastery for 2 rubles per pud, and they kept the money "all to themselves" (53, p. 103).

Returning to the story of Gvozdev: As soon as the Eskimo disappeared on the island, the most experienced servitors, E. Permiakov, L. Poliakov, F. Paranchin and A. Malyshev, again addressed the geodesist about returning because of "the lateness of the season," inadequate food (there was only a small amount of food) and a leak in the ship. The entire crew supported the request to return to Kamchatka. The sailors, seafarers and servitors wrote a petition "by their own hands" and gave it to Gvozdev and Fedorov. The geodesist and the assistant navigator "by mutual agreement" made the important decision. According to Skurikhin's testimony, bad weather broke the mast of the ship on the return voyage. With great difficulty on September 27 (28), 1732, the *St. Gabriel* reached the mouth of the Kamchatka River.

In all his reports (June 22, 1733, April 13, 1741 and September 1, 1743), Gvozdev emphasized that "on the sea voyage" he went "together" with Fedorov at the direction of Pavlutski. Since the Captain's order of July 13, 1732 has yet to be located, it remains a question as to what role the sick assistant navigator played. According to plan, on a sea voyage only one has authority of command. Though Fedorov tried several times to perform his duties, he often missed his watch. Regarding the determination of the course of the ship, all the servitors and sailors obeyed Gvozdev, based on Moshkov's statements.

Upon returning to Nizhnekamchatsk the seafarers began to prepare their reports. On December 19, 1732 Gvozdev sent a detailed report of the voyage "and necessary descriptions" to Pavlutski by special delivery through Ivan Soldatov to Anadyr. The information was forwarded to Yakutsk since the Captain had already left Anadyr. The geodesist reported "in what places and what number of islands they saw and what islands they visited and on what islands there were people and about the complete voyage...also about the supplies used during the voyage." Attached was a copy of the ship's journal (or lakhbukhov) "of the course from the mouth of the Kamchatka River to the islands and on the return to Kamchatka and about the leak in the ship" (30, p. 237).

Gvozdev did not prepare a map of the sea route of the voyage, because during the first 24 hours of the voyage Fedorov would not allow him to fill in the required information in the ship's journal and during this time it wasn't filled in at all. In addition, during the voyage "on many of his watches Fedorov didn't maintain the register." In Nizhnekamchatsk, on November 10, 1732 Gvozdev forwarded a letter to the assistant navigator suggesting they "mutually" complete the journal and the map of "the places we have visited and seen." However, Fedorov "wouldn't let" Gvozdev either correct the journal or prepare a map. The general tone of the responding letter of November 28 is reminiscent of the proverb, "Every cricket should know its perch."

In a haughty manner the assistant navigator commented that he was dispatched by the Admiralty Board not to design maps but for his navigational duties, for sailing and designing marine maps. After several "proud statements, "Fedorov further advised the geodesist "to mind his job and do his assignment and make maps, as he should." With regret, Gvozdev stated that this attitude toward the preparation of the reports deprived him of the possibility of using the ship's journal to prepare a map.

On June 22, 1733 five months after the death of Fedorov, Gvozdev prepared for "examination" by the office of the Okhotsk port the report of the voyage that was forwarded to G.G. Skorniakov-Pisarev from Kamchatka on December 23. Gvozdev sent the original ship's journal with the report, regarding which, as the geodesist wrote, "it was not possible to design a map." It is strange that neither Pavlutski, nor the Okhotsk office informed the Admiralty Board, the Tobolsk Governor or the Irkutsk Provincial Office about the results of the voyage to the Big Land. Most likely, at the beginning, neither Gvozdev nor the local authorities paid too much attention to the results of this ordinary voyage. Nevertheless, rumors about the voyage spread quickly. In 1736, V.I. Kazantsev, who visited Kamchatka in 1732-33, presented Tobolsk a working draft and notes about the voyage of the ship *St. Gabriel* to America. This draft has never been found. The Captain-Lieutenant relayed a story about the voyage: "We heard that the crew of the ship saw many foreign yurts made from dried skins on the cape opposite Chukotka, but they themselves did not go ashore. Between those two capes were three islands that are depicted on the working drawings. They visited two islands and saw foreign dwellings, yurts made in the earth. There were some residents but when they saw the ship they sailed in their boats to the American cape. From those small islands they sailed to a third big island. It was quiet there. There was no wind of any kind. Local natives arrived from that island in their boats. Only they didn't come very close to the ship. They shouted from a distance. When asked if it was possible to sail behind that big island along to the east next to the American cape, they advised them not to go; they said it was not deep. They were spoken to in the Koryak language and sign language."[15]

Several days after returning from the voyage, Gvozdev arranged for the sick assistant navigator to live in town. On October 3, Fedorov moved into the quarters of soldier Ivan Katashetsov. But the apartment didn't bring peace. On the contrary, the last months of the assistant navigator's life were troubled. On February 4, 1733 a week before his death, Fedorov wrote to Gens about his misfortunes.[16] The landlord of the apartment, for unknown reasons, didn't like his tenant, often scolded him and attempted "to kick him out of the apartment by force."

Fedorov was severely ill, helpless and harassed by his landlord. He was a pitiful figure, not the heroic personage portrayed by earlier scholars.

Gvozdev's account of his Kamchatka period is very short and restrained. In his petition of 1758 about his retirement, he recalled that in 1731, together with Gens, he departed from Bolsheretsk on a sea voyage "for exploration and creation of a map of the land opposite Cape Anadyr called The Big Land. Also, I was sent to Anadyr in 1732 for exploration of the islands and made descriptions and many reports to Captain Pavlutski" (52, p. 155). In 1738, at the request of the Admiralty Board, Siberian authorities in Tobolsk, Irkutsk, Yakutsk and Okhotsk conducted a fruitless search for the reports of the Shestakov-Pavlutski Expedition. Gvozdev briefly mentioned the reason for the missing documents. He stated that during all their "voyages and the Okhotsk trip, the navigator did not make any descriptions, map, or keep the ship's journal because of his eye illness and blindness, and the assistant navigator could-not because of his foot infection from which he passed away and he, Gvozdev, found that it was impossible to make them all by himself" (25, p. 45).

Footnotes to Chapter 3

1. TsGAVMF, Case 3, pp. 54,89; Case 5, pp. 150 and other side 167.
2. Ibid., Case 3, p. 57.
3. Ibid., Case 3, pp. 91 and the other side of 91.
4. Ibid., Case 5, p. 126-127.
5. Ibid., Case 4, p. 4 and the other side.
6. Ibid., Case 3, p, 188-195.
7. TsGADA, f. 1002, Case 1, pp, 5, 8, 9, 12-16, 18-20.
8. TsGAVMF, f. 216, Case 4, p. 61.
9. Ibid., p. 23 and the other side.
10. Here and further is presented the voyage of 1732 according to Gvozdev on September 1, 1743, preserved in TsGAVMF, f. 216, Case 53, p. 733-738 and other side). The same report with some errors was published by A.P. Sokolov (59, pp. 88-103) by A.V. Efimov (30, pp. 244-9) and in English (75, pp. 160-162; 76, volume 1, pp. 22-4)
11. The ancient Eskimo legend about the man who angered the gods told about the remote time when the shores of Asia and Alaska were connected by a long pebbled spit with only two - mountains, Imaklik and Inatlik. On the spit lived the sea hunters who worshiped the gods and didn't think that they were stronger than the gods. But there appeared one person who thought that man was a god himself, his own boss. One day during hunting, the god appeared to him in the form of a seal and began to talk to him in a human voice and began to admonish the hunter. But the hunter wouldn't listen and killed the seal and took its skin. Returning to the shore, he began to boast of his lucky hunting. By evening the sky darkened. A strong wind began to blow. The sea was restless and in the night there was a terrible storm. The sea, the sky and the land mixed together. The people ran to the two mountains to save themselves from the severe storm. By sunrise, everything was quiet and to the surprise of the hunters, there was a new surface. The pebbled spit had disappeared and between

the two mountains were only waves of the ocean (Rytkhow Yu. *Poliarny Krug*. Magadan, 1977, pp. 181-2).

12 Instead of a southwest route of the St. *Gabriel* from Alaska King Island, Danilov indicated the route to the northwest (24, p. 82).
13 TsGAVMF, f. 172, Case 408, part 1, p. 55.
14 TsGADA, f. 248, Bk 7312, p. 50 and the other side.
15 Ibid., the other side of p. 50 and p. 51
16 TsGAVMF, f 216, Case 4, p. 93 and the other side.

Chapter 4
The Commander of the Kamchatka Detachment

Based on the brief information Gvozdev provided about his stay in Kamchatka, it is known that in 1733, G.G. Skorniakov-Pisarev, commander of Port Okhotsk, directed navigator Gens and the servitors to Okhotsk. Gvozdev, the geodesist, was to remain in Nizhnekamchatsk for reconstruction of the ostrog. He remained there through 1735. There are few references to Skorniakov-Pisarev or Major V.F. Merlin and D.I. Pavlutski who governed Kamchatka. Gvozdev maintained a relationship with all of them for an extended period of time. Merlin and Pavlutski had unlimited power and wide authority in Kamchatka from their office in Okhotsk.

After the voyage to the Big Land, the Kamchatka Detachment slowly began to lose its independence. Pavlutski left Anadyr for a new appointment in Yakutsk, being replaced as leader of the expedition. Skorniakov-Pisarev appeared earlier as the new commander of Port Okhotsk and Kamchatka was under his authority. In a lengthy instruction from the Siberian office, he was to take energetic measures with the native population in the regions of Okhotsk, Yudoma Cross, Ud and Kamchatka for the accommodation of 300 servitors in Okhotsk. He was to bring horses, cattle and all kinds of bread grains to Okhotsk (50, volume VIII, No. 5753, 5813). Three navigators and six sailors transferred to Skorniakov-Pisarev's command for sea voyages along with former officer V.I. Kazantsev. The commander of Port Okhotsk was also responsible for overseeing construction personnel on Kamchatka. Pavlutski and his expedition were to report directly to Skorniakov-Pisarev.[1] The new commander of the port assigned Kazantsev to head the reconstruction and repair of the ostrogs on Kamchatka. Kazantsev, knowledgeable in the fundamentals of fortification, built Nizhnekamchatsk and Bolsheretsk from working drawings. Skorniakov-Pisarev decided to relocate Nizhnekamchatsk because of the presence of volcanic activity and many earthquakes at its previous site. As for Bolsheretsk, it was "remote" from the sea.[2]

Gregory Gregorievich Skorniakov-Pisarev was once an associate of Peter the Great. He had also been president of the Naval

Academy and prosecutor of the Senate. His participation in a plot against A.D. Menshikov resulted in his demotion and exile to Zhigansk. In Siberia, he drank a lot, constantly criticized everything and often quarreled with Yakutsk officials, Thadoviski, Pavlutski, Bering, Spanberg and others.

While in Yakutsk, arguments between Skorniakov-Pisarev and Pavlutski seemed impossible to settle, but the subordinates of the Commander of Port Okhotsk on Kamchatka carried out his instructions to the letter. On his return from the voyage, Gvozdev was the head of the ship and the Kamchatka Detachment. He found himself in difficult circumstances since he reported to three people; Pavlutski, Skorniakov-Pisarev and Merlin. This period of the geodesist's life, 1733 to 1735, was not previously studied.

On Kamchatka, Gvozdev followed the order he received on March 29, 1733 from Skorniakov-Pisarev. He was to immediately dispatch Gens to Okhotsk on the *St. Gabriel*. On April 6 Gvozdev sent a copy of the order to the navigator with the postscript: "and the ship *Gabriel* does not have so much as a small leak."[3] On the following day Gens demanded possession of the ship, all the supplies and materials for maintenance of the ship and sails. On April 20, Gens again addressed Gvozdev but did not receive an answer.[4] Gens had not worked on a ship for about a year, but knew very well that if he received the ship with a crew and equipment he would once again be back in a position of power and have a certain degree of independence.

His impatience and persistence in speeding up the events are understandable. In the meantime, relations between the geodesist and navigator continued to be tense. Gens was looking for a mediator to negotiate with Gvozdev. Kazantsev turned out to be an unexpected ally. It seems that Kazantsev also had a falling-out with Gvozdev and Speshnev. In his complaint of May 19, to Kazantsev, Gens presented the situation in detail. Gens was ordered to go to Okhotsk immediately but the geodesist would not turn over the ship and crew. "All these," contended the navigator, "are Gvozdev's improper deeds. Because of his imprudence, he creates obstacles in order to delay my departure."[5] There were rumors concerning the intentions of the geodesist about the maintenance of the ship when he started working on the keel. "And I know that Gvozdev doesn't understand this type of work and he might damage the ship even more,"[6] objected Gens. The navigator asked Kazantsev in writing to inform Okhotsk and to point out Gvozdev's "obstinacies and disobedience." At last, on May 22, Gvozdev spoke to Gens about "the return" of the ship, the supplies and the materials, "everything present" and also of the sailors, seafarers, interpreter and servitors. Regarding the delay in the transfer of the servitors, Gvozdev explained that they were all under the supervision of Speshnev and were sent for laying-in wood and boards for the repair of the ship. Also, some were sent to smoke, for resin and laying-in of dry wood and had not returned until now.[7] On May 23, Gvozdev sent Gens the crew of the ship and in the file he itemized the personal equipment of all the servitors.[8]

In May and June of 1733, Gvozdev and Speshnev worked towards preparing the ship for the voyage, organizing repairs and deciding who would be sent to Okhotsk first. On June 4, Gvozdev sent Gens a large quantity of writing paper and two pounds of thread for repairing the sails. On June 18, because of the shortage of food on the ship, he sent three sacks of rye flour with sailor A. Nazarev.[9]

The passenger list for the ship was complete. At the request of commissar I. Averstev of Verkhnekamchatsk, Gvozdev and Gens decided to use the ship to transport the construction assistants of Nizhnekamchatsk, I. Krikov, three servitors and O. Stupin with the treasury to Okhotsk.[10] Unexpectedly, Kazantsev wrote to the commander of the ship *Gabriel*, navigator Gens, telling of his intentions to depart for Okhotsk. In order to understand his motives, one needs to examine the previous report[11] from Gens dated June 2, 1733 to S. Gardebol.[12] According to the navigator, on May 31, while working on the ship, he went ashore to examine the pulleys. Kazantsev also happened to be there. At this time sailor Leonti Petrov informed them that Gvozdev kept him starving under security guard. Gvozdev's ordered him not let out of the tax house even for a short period of time, except for physical body needs. "I reported him on an important matter and now I'm dying of starvation, because he doesn't feed me or let me go to town."

Gens' attitude toward Gvozdev is known. Here he couldn't help but once again blame the geodesist saying, "It is very suspicious that the defendant keeps the plaintiff under security guard." Then the navigator recalled how, in May of 1732, "Gvozdev did not receive the sailors of the crew by written order, he took them by force." In his desire to defame Gvozdev, Gens falsified the facts. In reality at the beginning of May 1732, Gens had been removed from leadership of the detachment. Because of his illness, he was left in town. Turning over the ship, the people, and all supplies and property to Gvozdev, Gens wrote, "I'm transferring under receipt to you L. Petrov. He is in chains. His strength is necessary for the maintenance of the ship. He should be delivered to Okhotsk authorities. He is useful and very knowledgeable as a sailor."[13]

The navigator threatened to report, "Whoever it is necessary" if the sailor was not sent to work immediately. Gvozdev and Speshnev, however, ignored the threats and didn't give the sailor to Gens. From the beginning Petrov was kept "chained" by Gvozdev because he threatened the geodesist. Then Speshnev received custody of him in the coachmen's yard and kept him under security guard, "chained until his departure" to Okhotsk.[14]

On May 5, Kazantsev spoke out against Gvozdev, informing Gens about his remaining on Kamchatka to build two ostrogs on the Bolshaia and Kamchatka rivers. "To help with this work, I will need either geodesist Mikhail Gvozdev or the assay master Simon Gardebol." The plan of the construction of the ostrogs included using local Cossacks of the town, Cossack children, industrial people and the servitors in the detachment on Kamchatka.[15] Gvozdev was to send the servitors of the detachment for work.[16]

Upon his arrival at Bolsheretsk, Kazantsev said that on December 7, 1732 and again on March 15, 1733 he requested Gvozdev in writing to report on the work of the "Nizhneostrozhnye and Partovskimi" detachments of servitors who were woodcutting and constructing towers. On May 15, Kazantsev arrived at Nizhnekamchatsk and discovered that nothing was done and Gvozdev still had the people in his detachment. According to the information furnished by Kazantsev on May 28, the servitors forwarded a petition stating they didn't have any food or anything to work with and the place selected to build the ostrog was inappropriate, indicating that they found a more suitable location. Kazantsev complained that the servitors were allowed to go hunting for indefinite periods of time. When he complained, they began shouting at him that they knew better than he how long it takes them to obtain food for themselves. As a result of this the builder stated, "Half the summer passed, but no construction was done."[17] Further, Kazantsev wrote about his concerns. Discovering that Gvozdev was under suspicion and there was a serious charge against him, the excaptain-lieutenant recommended that Gardebol "separate Gvozdev from the crew and send him to Okhotsk." Under regulations, people under suspicion were not to be in charge of anything unless their innocence could be proved. On June 5, Gardebol left for Verkhni. "And" continues Kazantsev, "with him," Gvozdev. "I am concerned about the suspicion and complaints about him brought to me in writing. Besides this, craftman assistant, Ivan Speshnev, calls me a thief and uses obscenities in speaking to me in public." Kazantsev saw salvation from living in leisure and the threat to become "prematurely dead because of vicious accusations and such indecent deeds," in his urgent departure to Okhotsk.

On June 17, Gvozdev sent Speshnev to the ship with the public money and sable for the treasury, convict Petrov in stocks, and two security men. He asked Gens for a cabin in which he could keep the treasury and the file on the investigation and to inform him of the date of the ship's sailing.[18] On June 19, the date Gens had scheduled for the departure, Gvozdev informed the navigator that his final reports were unfinished. Most likely it was in this period when the geodesist wrote the 1732 voyage report, which has yet to be found. "And this day, please, don't leave but wait. As soon as I finish the reports you can go."[19]

On June 22, Gvozdev also sent to Gens on the ship another convict to report to the authorities, servitor Ilia Skurikhin, one of the participants of the voyage to Alaska. On June 17 Skurikhin had been shouting in the yard that he knew "important things" about boat craftsman assistant Ivan Speshnev.[20] The geodesist's report to Skorniakov-Pisarev contains the essence of the case, which was recently found, in addition to the accusations of Petrov against Gvozdev.

On July 9, the *St. Gabriel* left for sea from the mouth of the Kamchatka River. But the ship was poorly prepared for the hard voyage around Kamchatka and across the Okhotsk Sea. The Captain requested help from Gvozdev, demanding that he send carpenters with Speshnev and the repairmen. Gens was constantly dependent on the geodesist because the members of the crew, including the carpenters, and public property were under his supervision. It was not in vain that on July 11 he wrote to Gvozdev, "You ordered me through soldier Alexander Imiev, not to let people off the ship to go to the island because you were worried. But we cannot survive without dispatching people because of our needs for food. People have always gone ashore in such circumstances."[21]

Regardless of how few servitors there were, the majority of which were sent fishing, Gvozdev managed to gather the carpenters and sent them to the ship with Speshnev. The food and the lack of food supplies concerned the residents of Kamchatka, Russian and local. Thus, in the letter to Gens on July 5, Gvozdev emphasized that he left the island for 24 hours because of the extreme need for food. Even the runaway convicts from Olutorsk could go to the ship for food because of the shortages on Kamchatka.

In the middle of July the geodesist, the craftsman assistant and the navigator were busy repairing the *St. Gabriel* and selecting supplies, glass, compasses and the like. Regarding Gens' requests to send the extra carpenters to Speshnev's crew, Gvozdev expressed concern that Gens might take them with him on the ship creating obstacles to the construction work on ostrogs. Gvozdev, who was responsible for the construction, was not in a hurry to transfer the experienced carpenters for maintenance work. Preparation for the voyage took all July, August and half of September 1733. The work of the carpenters and blacksmiths turned out to be especially difficult during these repairs.

In the summer of 1733 the *Fortuna* brought important instructions from Skorniakov-Pisarev to Bolsheretsk. Through those instructions he tried to regulate the activities of the Kamchatka leaders. Most of the decree instructed Gvozdev and Speshnev to "govern the residents of all three Kamchatka ostrogs including the trade people, industrial people and the servitors. They were also to take a census."[22] On July 23 Gvozdev and Speshnev wrote to Gens "now one of us is going to the ostrogs with a crew to work."[23] On July 29, Speshnev informed the navigator about his departure from Nizhnekamchatsk to take over Bolsheretsk and Verkhni from the commissar, I. Everstov, and to send the commissar away. "And the geodesist Gvozdev, who I ordered to come here," continued the new authority of the town, "will remain here until the work is complete."[24]

So, the commanders of the detachment, geodesist Gvozdev and craftsman assistant Speshnev, began to govern Kamchatka. In their charge were all the Kamchatka towns and all the servitors on the Peninsula. Gvozdev supervised the construction and reinforcement of the towns while Speshnev oversaw the maintenance of the ships. Gens strictly followed the formal rules of protocol. All his memos and letters were respectfully addressed either to Nizhnekamchatsk, "to the powerful sir, geodesist Gvozdev" or to Bolsheretsk, "to the powerful Speshnev."[25]

The major task of the Kamchatka leaders was to complete the summer instructions from Okhotsk. First, they decided to examine all "the local port employees and distribute 60 people to each town by mutual agreement."[26] It is necessary to note here that during the second half of 1733, the Kamchatka Detachment of the Shestakov-Pavlutski Expedition was practically disbanded. Its members were sent to the towns for guard duty and construction. Some of the servitors were sent to Okhotsk and Yakutsk. A small group was assigned the crews of the *St. Gabriel* and *Fortuna*.

On July 23, Gvozdev gave Gens the results of the examinations of the port employees for filing. There were 91 people registered on the list.[27] After examination, 60 people were sent to Gvozdev's crew. Speshnev brought 49 people for distribution to the ostrogs. With the transfer of nine more people to Speshnev at

Bolsheretsk on September 5, the formation of the Kamchatka garrisons was complete.[28]

The departure of *St. Gabriel* from Nizhnekamchatsk to Okhotsk became a special event because it affected, without exception, the entire Russian population of Kamchatka. In addition to the crew, all the servitors were sent on the same ship. These were the administrative people responsible for enforcing the regulations in the towns. Since 1731, they were in charge of collecting tax. There were several "secret informers" and people charged with violations of the law. Also sent were the public property, documents, reports and the runaway convicts from Olutorsk. The story of their appearance on Kamchatka is interesting. In the summer of 1732, Gvozdev questioned eight servitors who had left Olutorsk because of starvation and appeared in his detachment. Their statements and his report were sent to Anadyr with I. Soldatov. On October 17, 1732 Pavlutski ordered the geodesist "to search for runaways and keep them under strict security guard once they are caught." This order was brought to Gvozdev from Anadyr on January 21, 1733 by a friend of Petrov, T. Krupyshev. Together they had been the leaders of the detachment in Tauisk during the winter of 1729-30. The geodesist sent Krupyshev with his wife and children to Olutorsk "because the town was newly built and should be populated."[29]

On January 23, 1733 Krupyshev told Gvozdev that during his journey with the reports from Nizhnekamchatsk to the Anadyr and return, Petrov traded a copper pot for 30 red foxes, 2 sacks of sweet herb for 20 foxes, and the rest of his belongings. And, "while residing in my house, the sailor sold all the dogs, used all my fish and bankrupted me." Although questioned because of this report, Petrov would not give a signed statement. Unexpectedly, as later explained by the geodesist, he "reported me to the tax department because he was sent to Okhotsk."[30]

By the end of July the ship dispatched by I. Everstov arrived for the tax collected by the tax collectors and the servitors.

By the beginning of September, the tasks set out by Skorniakov-Pisarev to the Kamchatka administration were completed. Geodesist Gvozdev and apprentice Speshnev were in charge of 60 people and 8 sick men left in Kamchatka. Navigator Gens on the *St. Gabriel* had 48 sailors, seafarers, soldiers and servitors.[31] After repairs and loading of people and furs, the *St. Gabriel* headed towards the mouth of the Bolshaia River. From Bolsheretsk, Gens took on board I. Everstov, M. Borisov, A. Shtinnikov and all the servitors who were to be taken to Okhotsk. The voyage ended well. On September 25, 1733 *St. Gabriel* arrived in Port Okhotsk. The same day V.F. Merlin demanded from Gens a detailed report about those who had arrived with him and what they brought from Kamchatka. Also demanded was information about food reserves for the ship's crew and servitors.[32]

The *St. Gabriel* remained at Okhotsk for 10 days. On October 2, under the order of the Investigation Department of the Okhotsk Office, Gens was to transport five puds of lead to Kamchatka. On October 5, 1733 the *St. Gabriel* left Okhotsk port. Sailing with Gens and the crew to Bolsheretsk were Major V.F. Merlin, the newly appointed commander of Kamchatka; a nobleman from Irkutsk, I. Dobrynski; and also "merchants and industrial men." After reaching the ostrog with great difficulty, Gens worked to prepare the ship for the following summer's voyage. He asked for help on November 4 and December 3 from Merlin, trying to get instructions to Gvozdev to send servitors and supplies. For the repair work, Gens requested four carpenters, a blacksmith, 3 puds of iron, 100 arshin of thick canvas and 4 pounds of thread. They already had the rest in Nizhnekamchatsk with Gvozdev. The same day, the navigator sent sailor L. Smetanin to the geodesist. He was to return to Bolsheretsk with the rest of the material no later than April, 1734.

One of Merlin's first decisions in Kamchatka was the order of November 30, 1733: navigator Gens was to immediately receive all "port" servitors and all the supplies and materials from Speshnev.[33] In addition, Merlin ordered Speshnev to go with Dobrynski to Verkhni and Nizhnekamchatsk to bring "the treasury" to the new Kamchatka commander. Most likely, the Major discovered errors by Speshnev and decided to substitute the sailor's report to the new leaders. After receiving the order from the head of the investigation office, Speshnev was informed on December 2 about the possibility of transferring only 19 servitors to the detachment at Bolsheretsk.

He tried to explain the difficult circumstances to Gens; "It is impossible for me to determine if these servitors have any ammunition or any kinds of supplies since I was not the only one in charge of this detachment. Besides me, there is geodesist Mikhail Gvozdev who is in Nizhnekamchatsk. That is why the requested crew is not transferred to you without Gvozdev."[34]

With Merlin's arrival, significant changes took place in the management of Kamchatka and the detachment. Gens continued to command the ship *St. Gabriel* and accepted the Bolsheretsk crew to the detachment. The apprentice and the geodesist transferred Kamchatka to I. Dobrynski and began to fulfill their major responsibilities. Speshnev worked preparing lumber and repairing the ship. Gardebol was involved in the search for ore and minerals. Gvozdev continued to be in charge of the reinforcement of the ostrogs. Under his direction were the carpenters, blacksmiths and other specialists. Skorniakov-Pisarev also made him responsible for one of his projects, sending a large detachment to build a road and establish a post office on Kamchatka.

On June 7, 1734 Merlin ordered Gens to go to Okhotsk and bring food and Major Pavlutski on the return voyage. On August 13, the lumber collected in 1733-1734 was brought to the ship. By November of the same year, Gens returned to Kamchatka. He made a complete inventory of the rigging, materials and supplies as requested by Pavlutski.

The year 1735 arrived, a year of crucial changes in the destinies of Gvozdev, Speshnev and Gens. Because of the investigation into reports by Petrov and Skurikhin in the second half of 1735, Gvozdev, Speshnev and Gens left Kamchatka. In a special file of the detachment of the Okhotsk department they, sailor Petrov and navigator S. Khitrov, went to Irkutsk "on an important case." V. Bering prepared this file for the Admiralty Board May 10, 1737.

They had commissioned Bering not only for the leadership of the Second Kamchatka Expedition but also for supervision of the establishment of the Okhotsk port. The file contained the names of 29 sea and admiralty servitors who had been dispatched to Skorniakov-Pisarev (69, pp. 77-78).

On August 4, 1735 the *St. Gabriel* again arrived in Port Okhotsk. An assistant of V. Bering, Captain M.P. Spanberg, was staying in Okhotsk and was in charge of shipbuilding and construction of residential and administrative buildings for the Second Kamchatka Expedition. Spanberg directed Gens to give all the sailors provisions for the month of August and "appoint" work for them. Fourteen of the servitors who came on the ship were sent to the Okhotsk tax office. After this, the *St. Gabriel* provided the expedition with ship supplies that were delivered to assistant constable Uri Aretlander and shipman Stephan Serebriakov by L. Smetanin.[35]

On September 9, Spanberg received a letter from Skorniakov-Pisarev suggesting he send navigator Gens to the office of Port Okhotsk, chained and under the strong security, based upon a dispatch from Tobolsk of an important secret case that came through the Irkutsk Provincial Office. The next day, Sergeant M. Kuznetsov of the Kamchatka Expedition handed Gens over to A. Tuev.[36] On March 29, 1736, Gens was sent to Tobolsk where he remained on bail. In his last written report in September of 1737, not long before his death, the navigator was going through a lot of hardship and requested the Siberian Governor's office to issue him a salary for May through September. Several years later the Admiralty Office issued the retained salary to his widow Maria, because of her poverty and small children.[37]

The investigation of Speshnev proved justified. On October 30, 1739 the Siberian office ordered him to pay the Admiralty Board 5 rubles, 30 kopecks for 2 carriages and confiscated all his contraband.[38]

On May 20, 1737 the Siberian Department examined the formal reply of P. Buturlin, the Tobolsk Governor, about Gvozdev, with a copy of the report from Vice Governor A. Plesheev of Irkutsk attached. He in turn presented the contents of the report of the Yakutsk office. A. Laborovski of Yakutsk reported the arrival of the geodesist. He was on the way to Irkutsk because of the investigation. On September 26, 1735 the Yakutsk office obtained all public property from Gvozdev. The geodesist explained that he was with a detachment to develop land maps and that he was given the instruments in St. Petersburg by the Academy. He said there was a file in Moscow containing information on the materials he had. The instruments and materials were taken from Gvozdev in Yakutsk for preservation until the decree.[39] The Irkutsk Provincial office indicated that all the items be preserved and not used. As it turned out, neither the geodesist nor Tobolsk had any idea about the investigation by the Siberian Department. That is why the information was sent to the Irkutsk Vice-Governor and why Gvozdev was sent from Yakutsk to Irkutsk. In the journal of the meetings of the department appeared the note: "If the geodesist is released, the instruments and materials should be returned to him. If he, Gvozdev is not released then instruments and the materials should be given to the detachment he, Gvozdev, was with."[40]

While the Siberian Department was deciding about the public property the geodesist left in Yakutsk in 1735, Tobolsk completed its investigation into the matter of accusations of state crimes against Gens and Gvozdev. During the interrogation at the Siberian Governor's Office and Secret Investigation Office, Petrov admitted that he didn't know the important things he said he did about Gens and Gvozdev and didn't hear about it from anyone and that he started it all in vain.[41]

The Secret Investigation Office sent Petrov to the Moscow Admiralty Office for determination of punishment in accordance with the law of false evidence. At the same time, they also determined that Gens and Gvozdev had nothing of importance and found them not guilty.[42] Based on a decree of October 11, 1733 Petrov was punished for making a false report by walking five times through a line of 500 soldiers who beat him with rods. On November 20, 1736 he was sent to the Admiralty Board for decision regarding his future service and from there on December 16 of the same year he was sent to Krondstadt.

On February 14, 1738 the Admiralty Board studied the information from the Secret Investigation Office. They decided to issue an order on behalf of navigator Gens and geodesist Gvozdev, "at their current locations." It stipulated that both of them, if they were kept under security guard, should be released, their salaries paid and Gens be dispatched as before to the Kamchatka expedition to Captain-Commander Bering. The order directed the geodesist Gvozdev be found and sent to his previous assignment (69, p. 74-79).

The Admiralty Board, which decided to issue the salary of the deceased Gens to his widow (he passed away in 1737), couldn't find out for five years if the navigator had any service debts. That's why in the decree of September 7, 1742 forwarded to both Bering and Spanberg, the Board warned that if in case of the discovery of any debts of the navigator or any unauthorized expenditure of public property, they would penalize those who sent no reports after many years. Spanberg noted in his response to previous requests for information about the bills of Gens, the absence of the required information in the Expedition's files. Pavlutski and Skorniakov-Pisarev didn't forward any answers at all. The navigator had served in the detachments of both.

It is not known how Gens was registered with A.F. Shestakov or under what conditions. The report of the geodesist Mikhail Gvozdev speaks of a decree in 1730 that Gens received the ship *Gabriel* and supplies from the Yakutsk boyar son, Ivan Shestakov. Gvozdev did not know what type of supplies Gens received with the ship. By order of Pavlutski as directed by the Okhotsk department, Gvozdev returned the ship and supplies to Gens in 1733. After receiving it, Gens transported the tax treasury to Okhotsk and returned to Kamchatka. After that the ship was turned over to Spanberg.[43] The Board was concerned about not being informed of all these operations. Spanberg received the decree in Yakutsk on May 23, 1743. On November 5, he verified the absence of information about the navigator "because of the fact that Gens was never in my detachment." He reported that he "thoroughly" informed V. Bering about receiving the *St. Gabriel* in 1740.[44]

Gvozdev briefly described the events of 1735 to 1738 in a biographical petition of 1743: "In 1735, because of the statement made against me by the sailor Leonti Petrov, I was taken to Tobolsk. I was kept in the Siberian Provincial Office through July of

1738. The decision in this matter exonerated me" (25, pp. 43-44). The emphasis placed by the geodesist in a 1758 document is somewhat different: "I was in the Kamchatka Detachment through 1735, the year of the arrival of Pavlutski. Because of the secret, false report about me, I was sent to the Tobolsk Provincial Office. While I was innocent, I was kept there through July 1738 when I was released" (52, p. 157).

Footnotes to Chapter 4

1 TsGADA, f. 248, Bk 164, p. 733.
2 Ibid., Bk 666, p. 443.
3 TsGAVMF, f. 216, Case 3, p. 219.
4 Ibid., p. 220; Case 4, p. 29 and the other side.
5 Ibid., Case 4, p. 3.
6 Ibid., the other side of p. 3.
7 Ibid., pp. 3, 15, and the other side.
8 Ibid., Case 4, p. 222.
9 Ibid., Case 3, pp. 208 and the other side;
10 Case 4, pp. 43 & 234; Case 5, p. 31.Ibid., Case 4, p. 36 & 207.
11 A document of official correspondence between peers.
12 TsGAVMF, Case 3, pp. 47 and the other side.
13 Ibid., pp. 187 and the other side.
14 Ibid., Case 4, p. 96.
15 Ibid., pp. 121 and the other side.
16 Ibid., p. 29.
17 Ibid., p. 121.
18 Ibid., Case 4, the other side of p. 203.
19 Ibid., p. 176.
20 Ibid., Case 3, pp. 216 and the other side.
21 Ibid., Case 5, pp. 2 and the other side.
22 Ibid., Case 4, p. 194.
23 Ibid., the other side of p. 52.
24 Ibid., pp. 195 and the other side.
25 Ibid., Case 3, p. 13; Case 4, pp. 42 and 67; Case 5, p. 8.
26 Ibid., Case 4, pp. 52-53.
27 Ibid., pp. 180 and the other side.
28 Ibid., Case 3, pp. 11, 14, 27 and the other side; Case 4 pp. 64 and 227; Case 5, pp. 8 and the other side.
29 Ibid., Case 3, p. 13; Case 4, p. 53.
30 Ibid., Case 4, pp. 40 and the other side.
31 Ibid., p. 62.
32 Ibid., Case 5, pp. 14 and the other side, 24 and the other side.
33 Ibid., Case 4, pp. 178 and 223.
34 Ibid., pp. 183 and the other side.

35 Ibid., Case 14, pp. 198-199; Case 99, p. 370.
36 Ibid., Case 14, pp. 9-10.
37 Ibid., Case 4, p. 177; Case 99, p. 370.
38 TsGADA, f. 24, Bk 14, Part 19, pp. 145 and the other side; BK 15, Part 1, p. 467.
39 Ibid., Bk 14, Part 10, .p iv (N236); Part 11, pp. 78 & other side
40 Ibid.
41 TsGAVMF, f. 212, Description 7, No. 206, p. 116.
42 Ibid.
43 Ibid., f. 216, Case 2, pp. 215-216.
44 Ibid., Case 99, p. 370.

Chapter 5
The Cartographer of Port Okhotsk and of the Second Kamchatka Expedition

In the summer of 1738 Gvozdev was sent from Tobolsk to the Irkutsk Provincial Office by decision of the Admiralty Board. In May of 1739, he was assigned to G.G. Skorniakov-Pisarev. "And upon my arrival at Okhotsk," noted Gvozdev, "I was in the detachment of Port Okhotsk and remained there through 1741 fulfilling my duties." By this time he had already served 12 years in Siberia. Deprived of promotions and salary increases, with no prospect of further promotion in service and mistreated, Gvozdev made an attempt to change his status in July 1740, two years after his reassignment. Through Bering he simultaneously sent a petition to Empress Anna Ivanovna and an application to the Admiralty Board. But he did not receive a response from either.

During the more than four years Gvozdev was under investigation, noticeable changes occurred in Okhotsk. 1737 saw the formation of the Okhotsk fleet of the Second Kamchatka Expedition. After restoration, Gens delivered the *St. Gabriel* to Spanberg in 1736, joining the *Fortuna*. Under the supervision of Spanberg and the leadership of the masters Rogachev and Kuzmin, the brigantine *Arkhangel Mikhail* and the three-masted *Nadezhda* were built between 1735 and 1737. A view of the four ships in Okhotsk is depicted in 1737 on "The Plan of the Okhotsk Ostrog."[1] Still, in 1736 Skorniakov-Pisarev began to move service buildings from the old ostrog to the port that he founded. However, because of severe rains and a flood at the mouth of the Okhota River, many buildings on the narrow island were damaged. The choice of location for the port and settlement was the subject of heated discussions and precipitated a quarrel between Skorniakov-Pisarev and Spanberg and Bering. Skorniakov-Pisarev refused to listen to arguments, insisting that there was no better location for the new town and shipyard than the one he selected. He accused Bering and Spanberg of "wanting the affairs of Port Okhotsk to be in chaos and to damage Pisarev by their shameless lies."[2]

By Senate decree of August 20, 1739 Anton Devier replaced Skorniakov-Pisarev as commander of Port Okhotsk. Devier received orders to examine "the low places in Okhotsk" and also determine "where it would be best to locate buildings." On February 28, 1740 the Siberian Department emphasized to the Siberian Administration and Devier that an examination was necessary in order to make a decision on the settlement's location.

The Okhotsk fleet of the Second Kamchatka Expedition in 1737.
A fragment of the plan.
(TsGADA, f. 192, Irkutsk Province, No. 104)

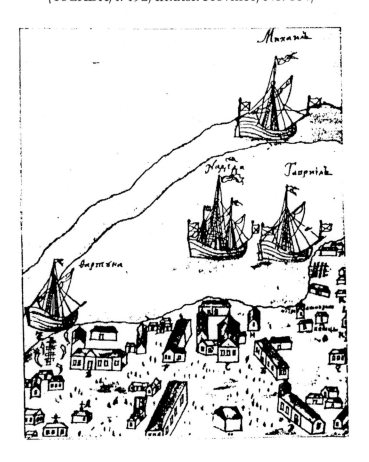

By the end of 1740, Devier examined a location suitable for the settlement and establishment of a shipyard, 30 versts up the Okhota River, at the junction with its tributary, the Malchikan River. Two delegations accompanied him during the examination. These representatives were from the Yakutsk office, and also "the best people" from Okhotsk, the geodesist, the priest, the ship-master, the sergeant, the quartermaster sergeant, three sub-warrant officers and five servitors. They determined that the town of Malchikan was suitable for the settlement and had much better environmental conditions than the place Pisarev had selected "near the sea on rotten rock." At Malchikan "there were forests, grass for hay, adequate fish and the water was fresh and didn't freeze in the wintertime." Moreover, the area was "not dangerous."[3]

In his report to the Admiralty Board about the advantages of Malchikan, Devier mentioned all those who had participated in the project. Attached to the report was the plan for the location that Gvozdev had developed. Fulfilling his duties, the geodesist helped Devier in the description of a suitable place for the settlement and development of a detailed working drawing. The Admiralty Board carefully considered the assumptions of the Okhotsk commander. Devier didn't indicate whether Malchikan was convenient for shipbuilding or whether ships could reach Okhotsk without hindrance. The Board recommended Devier leave the shipyard in its present location if the new one would create difficulties with shipbuilding and launching ships.[4]

In 1741, the reconnaissance and exploration of the Okhotsk Sea began. Captain William Walton helped complete the description of the shore to the north from Okhotsk to the Inia River (95 versts) and south to the Uliia River (144 versts). He was commander of one of the ships of Spanberg's detachment that visited Japan on two occasions. The Okhotsk office understood that Walton placed Gvozdev in charge of describing the shore to the south of Okhotsk. The Captain's instructions were to describe all rivers flowing into the sea within 200 versts of Okhotsk and to measure depth at the mouths of the rivers. They were to pay particular attention to the Uda River "to observe various forests and places for agricultural purposes" and to determine the possibility of mooring ships during the winter. They were also to describe all islands visible from shore. "Someone suggested that Gvozdev himself visit the islands located near the coast for better descriptions and observations." Bering ordered the geodesist be given one or two boats with necessary provisions so that he could cross rivers. He was also to present a journal and map of his voyage upon its completion (69, pp. 168-9). Gvozdev made some references to the completion of this task in his 1758 autobiography, including how he was dispatched "to describe the sea-coast from the mouth of the Okhota River towards Udsk, and that is what I described." This probably refers to the period from mid-April to July 14, 1741.

It wasn't clear to early historians researching Gvozdev's voyage to Alaska why the detained I.F. Skurikhin had made his report about the voyage to the Okhotsk office in April, 1741 (59, p. 82). One can see the reason for the interest in St. Petersburg, Tobolsk, Irkutsk and Okhotsk in the Chukotka Peninsula, the Big Land and the islands in the Bering Strait. On September 15, 1742 Lopez Lang, Vice Governor of Irkutsk, reported receiving a description of Chukotka life from Devier in Okhotsk. The Okhotsk commander sent the map and descriptions prepared by Ia. Lindernau to Irkutsk "for better knowledge and information about Chukchi life, over which areas and rivers they migrate, the closest Russian territory they occupy and the shortest route to them."[5] The practical value of such documents can easily be seen. The Okhotsk administration tried to obtain this same information from the people who participated in the 1732 expedition. The information was sure to be familiar to the author of the 1742 map used in describing the Chukotka coast, the islands of Bering Strait, and part of the west coast of Alaska. That is why Skurikhin's announcement to the Okhotsk administration was unusual. He participated in the Shestakov-Pavlutski Expedition and later made false accusations against Speshnev in Okhotsk.

Twelve references to the two Skurikhin reports were found. Nineteenth century historians used this information to describe the expedition to the Big Land: Sokolov published part of the text of the April 10 interrogation (51, p. 397; 59, pp. 104-107). Soviet scientists, especially A.I. Andreev and A.V. Efimov, found some additional materials: original notes of Skurikhin's story written on April 8 and mentioned in Devier's May 1, 1741 report to the Irkutsk administration; a copy of the full notes of April 10 that G.F. Müller received from Captain T. Shamlev; and a description of "the announcement" and brief story of the expedition as told to M.P. Spanberg by Afanacy Zybin on April 20, 1743 (53, p. 101-106; 30 p. 236-238). Spanberg added Zybin's notes to his report of November 5, 1743.[6] Lang reported Skurikhin's story on two occasions: in a report to the Siberian government on October 27, 1741, and to the Senate on September 15, 1741.[7] In the first report which the Siberian government received January 19, 1742, the Vice Governor summarized Skurikhin's story "about his being on the ship *Gabriel* in 1732 and being a member of the crew with Gvozdev and assistant navigator Fedorov on the voyage from the mouth of the Kamchatka River. He told of their making their way east from the mouth of the Anadyr River for five days when they saw land to their left. People lived on the land. They decided to bear to the left. They continued to move in that direction for five days before turning back. One native man visited their ship and invited them to come and see his land and offered them food. They didn't stop on the way back to Kamchatka. It would take five days to get to that land in good weather and about one to one and a half days from the Chukotka Peninsula."[8] The Okhotsk office compared the story told by Skurikhin on April 8 and 10, 1741 with the documents sent by Gvozdev on December 23, 1733. This report is still missing. Results of the comparison are not very impressive. Devier reported to Irkutsk that according to Skurikhin's report, the existence of the Big Land was evident (53, pp. 104-105). It was decided to obtain the necessary explanations and a map of the voyage from Gvozdev, the person who had played the primary role in the expedition. The commander of the Okhotsk port directed Gvozdev to complete this task.

As ordered, Gvozdev presented a new report about the expedition to the Big Land in April, 1741 according to various archives and historical sources. Only A. Polonski stated the exact date of the geodesist's report presented to the Administration on April 13 with S. Waxell present. In later publications of this document (by A. Zybin on April 20, 1743 for example), the exact date was not mentioned because the original report could not be located. According to Skurikhin's story, there were several versions of Gvozdev's report of April 13, 1741 and Zybin's variation is probably closest to the original. Devier's report to Irkutsk on May 1, 1741 and Lang's report of October 27, to the Siberian government, both mentioned only a small portion of Gvozdev's report about August 21, 1732 and the Big Land.

Devier suggested that the Irkutsk administration direct Bering or the people in Anadyr "to make small boats and find the land." The basis for this suggestion was the belief that the Big Land was very close to the Asian Coast (Cape Chukotka to the mouth of the Anadyr River is between 1.5 and 5 days by ship). Offering to research the unknown region, Devier mentioned that he was "quite sure that the Big Land was part of America." The Siberian government looked through the documents about the expedition and Lang's notes about Devier's report and finally, on February 1, 1742 adopted suggestions about "determining and finding the Big Land and islands" and proposed the Senate consider letting Bering continue the research.[9]

The geodesist was unable to complete the task for Devier. Gvozdev prepared a report but didn't present any maps describing the voyage. We can assume that he repeated all the arguments that it wasn't possible to complete maps without the main source of information, the ship's diary. In spite of this, Devier told his superiors about Gvozdev's failure to complete his task and about "his obstinacy towards the order." On July 6, 1742 the Irkutsk administration issued a new order that noted the "obstinacy" of the geodesist. Also, this order suggested interviewing all the officers of the expedition to get "accurate information" about the islands and the Big Land "and send it to Irkutsk." If this failed, the Okhotsk administration was "to build ships and organize an expedition in order to get the information and description of the Big Land and the islands."

In the report to the Senate, the Vice Governor of Irkutsk presented some ideas about the Chukchi and the expedition of 1732.

First, Lang assumed that the Chukchi of the Peninsula "had similarities and maintained friendly relations with the land which Skurikhin and geodesist Mikhail Gvozdev had seen" and that these unknown peoples helped the Chukchi and communicated with them. Lang was surprised that "officers of the expedition had already been in the areas and had made many different voyages but that this land and islands were still unknown with no information about them. I received indirect confirmation in the form of rumors that the land existed from various individuals."[10]

Confirming the necessity to organize a new expedition from Okhotsk to the Big Land and the islands, the Irkutsk administration established an agenda for Pavlutski, designated as the expedition's leader. While "there he was to pay careful attention and ask questions of the natives." He was to obtain various information: the distance from the home of the Chukchi to the islands and the Big Land; characteristics of communications possibilities; descriptions of the environs; information about the native peoples, their languages, political status, monetary system and raw materials. He was to find out about "any island or land still unknown and under which country's authority it existed."[11]

At Spanberg's initiative of July 14, 1741, Gvozdev joined the Second Kamchatka Expedition. We can assume that Devier, after the conflict about the report, did not object to transfer Gvozdev from the Okhotsk command. By that time, the fleet of four ships (the packet boat *St. Joann*, the *Arkhangel Mikhail*, the *Bolsheretsk* and the double-sloop *Nadezhda*) had returned from their voyage to Japan and was preparing for their next voyage. The commanders of the ships were Captain Spanberg, Midshipman A.E. Shelting, Midshipman V. Rtishev and M. Kozin. They were not able to complete preparations for the 1741 expedition until the end of the summer and missed the most suitable time for the voyage. Spanberg sent three ships to Bolsheretsk for the winter and directed Shelting and Gvozdev to continue research of the Okhotsk Sea. The journal of the double-sloop *Nadezhda* describes the voyage from Okhotsk to the Shantar Islands, the Kurile Islands and Kamchatka from September 4 to October 14, 1741.[12]

Sailing west from Okhotsk along the coast, the crew of the double-sloop *Nadezhda* saw the Shantar Islands on September 9. Gvozdev and navigator Rtishev made notes to this effect in the ship's journal. During the voyage, the members of the expedition took soundings, made meteorological notes and described the coast. On September 15, the *Nadezhda* entered the mouth of the Uda River and determined its location to be 55°30' north latitude (with an error of 48' to the north of the actual location). Gvozdev accompanied two boats to Udsk ostrog and back on September 21. Shelting summarized his voyage with the geodesist characterizing the geographic and hydrographic skills of the members of the expedition. In their reports they followed their instructions very closely, describing various matters. "In the river's mouth at ebb tide," the midshipman wrote, "The depth was 3 feet. During high tide, the depth was 12 feet, 4 inches. The Uda River has many mouths that are very shallow. The forest on the shores of the river is useless for shipbuilding" (52, p. 478).

Between September 22 and 28, the *Nadezhda* was near the Shantar Islands. They turned back because of a leak in the ship. On October 12, they had to stop because they were headed towards the Chekavka River at dusk.[13] On October 14 the *Nadezhda* joined the rest of the fleet at Bolsheretsk. On November 28, Shelting submitted a report of the voyage together with a map and the ship's journal.

Autumn storms and fog and leaks in the ship prevented Shelting and Gvozdev from completing their task. However, they did manage to finish their description of the mouth of the Uda River and the Shantar Islands. Spanberg supplemented his report of May 20, 1742 with copies of the instructions, the midshipman's report and some stories the native people had told Shelting and Gvozdev about "their land, forest and the Uda River."

Preparations for the new expedition to Japan were completed in Bolsheretsk by the spring of 1742. Spanberg appreciated Gvozdev's work in the expedition and sent him as an assistant navigator on the double-sloop *Nadezhda*. On May 2, 1742 the navigator of the ship, midshipman V.A. Rtishev, received word from the commander of the expedition concerning Gvozdev's promotion.

On May 23, the fleet of four ships with the *St. Joann* as the flagship began a voyage to the Kurile Islands but they encountered heavy seas and strong winds. First the ship *Arkhangel Mikhail* and the double-sloop *Nadezhda* turned back, followed by the *Bolsheretsk*. The *Nadezhda* reached 46°21' latitude before turning back.

Later, Spanberg recalled that on July 11 at latitude 50°39' sailors on the *St. Joann* saw the *Nadezhda* approaching the Kurile Islands. The two ships passed through the strait between Shumshu and Paramushir Islands where they met the other two ships of the fleet. "Then, in the report of December 11, 1742 the commander of the expedition told of sending midshipman Rtishev and midshipman Shelting on the *Nadezhda* to describe the Amur and Uda Rivers.[14] The new commander of the *Nadezhda*, Shelting, received orders on July 23. According to the directions of 1741, he continued the description and proceeded from the mouth of the Uda River to the Tugur River and then to Amur, to 46° north latitude. Being near Yudoma Cross and not having any information from the *Nadezhda*, Spanberg ordered Shelting to go to Yakutsk. "If he will not have many expenses and will have enough time, he was to bring Rtishev and Gvozdev with him and to bring all the maps and documents of the voyage." If he would not be able to leave, he was to remain in Okhotsk with his crew and send the map of the voyage to Spanberg with someone.[15] With his report of December 11, 1742 to the Admiralty College, Spanberg sent the map and journal of his voyage of 1742 and the journal of Shelting's voyage to the Uda River in 1741. The navigation route was shown on Spanberg's map of 1742.[16]

There is a journal of "the *Nadezhda* of the voyage of Shelting from the Kurile Island, Alaida, north latitude 50°40', compass declination ¾ in 1742." The journal contains the information from July 22 through October 13.[17]

On July 24, the *Nadezhda* headed west from Alaida Island and reached the eastcoast of Sakhalin near Terpeniya Cape. On August 3, the seafarers determined the geographic latitude of the ship. Gvozdev posted in the journal: "47°35' north latitude, on the coast we see a forest. The shore is low" (49, p. 75). From Terpeniya Gulf, the *Nadezhda* sailed north along the east coast of Sakhalin. At latitude 50°10' they turned and went south to 45°30' north latitude. Being near the east coast and Laperouse Strait, they could not complete their major goal of describing the coast which was six miles away from the ship and nearly impossible to see because of strong winds and heavy fog. On August 20, "because of the leaks in the ship and shortage of provisions," the double-sloop turned back and arrived in Okhotsk on September 10. Finally, on September 13, they "came into the Okhota River and left the ship in anchorage."[18] A week later, Shelting advised Spanberg of the return of the *Nadezhda* to Okhotsk with the map and the ship's journal.

During 1741-1742, Gvozdev not only carried out his duties as a geodesist but performed as a skilled navigator as well. His duties included: description of the coast; navigational measurements and astronomical research; keeping watch with the commander of the ship and navigator and maintaining the ship's journal. According to the journal notes, he worked on shore with Shelting and Rtishev to complete the final maps of the voyages. We cannot say that Gvozdev was in an extraordinary circumstance but, not having enough officers and well-trained specialists, geodesists often had to work in many different capacities. For example, Geodesist-lieutenant I. Svistunov sailed with Spanberg's crew to Japan in 1738-1739. In 1741 he was commander of the ship *Arkhangel Mikhail*. Geodesist-lieutenant P. Skobeltsin, in 1742, sailed under Spanberg's leadership on the packet boat *St. Joann*. In his autobiography of August 24, 1743 Gvozdev listed the work he did while serving in the Second Kamchatka Expedition: "I joined his crew and sailed on the double-sloop *Nadezhda* on a voyage with midshipman Shelting to describe the coast up to the mouth of the Uda River and the shore to the Amur River. In 1742, I was on the same ship with midshipman Basil Rtishev to write descriptions as far as the coast of Japan. In 1742, I was with Shelting on the same ship to describe Amur's shores and mouth. Being there, I completed my duties" (25, p. 44). In his biography of 1758, the geodesist repeated the same information but mentioned at the end one important detail: "that in many voyages, besides my direct responsibilities, I worked as a navigator" (52, p. 148). The new commander of Okhotsk reviewed Gvozdev's voyages during 1741 and tried to bring him back to Okhotsk in September 1742. On February 21, 1743 the Okhotsk secretary ordered Shelting to send Gvozdev to Okhotsk "for better use of his skills."[19] On February 24, the midshipman reported that "it wasn't possible to send Gvozdev" because of the preparations for the voyage to Yakutsk. On April 20, A. Zybin asked Spanberg a second time about Gvozdev. Spanberg declined his request, undoubtedly appreciating the professionalism, determination and the Far East experience of the geodesist. As a result, Gvozdev became a member of the Second Kamchatka Expedition after many voyages. But the activities of the expedition ended. It was 1743.

As directed, the geodesist performed the duties of an officer on many occasions but was not promoted for a long time. In this manner, he was equal with the officers and listed with them on various rosters and other documents of the expedition. For example, on July 18, 1743 in the Preobrazhenskaya Church in Okhotsk, 67 men swore their allegiance to Peter Fedorovich, cousin of Elizabeth, declared heir to the throne on November 7, 1742. The oath was signed in the usual manner, according to seniority as follows: midshipman A. Shelting, midshipman V. Rtishev, skipper D. Korostelev, navigator M. Petrov, navigator A. Shaganov, geodesist M. Gvozdev, constable I. Bobovsky, and the others by their own hand. The report of January 12, 1743 to Spanberg regarding soldier Mikhail Polujanovsky from the Yakutsk Regiment is interesting. Members of a special officer committee signed it with V. Rtishev as its head. Gvozdev is among the committee members. In another report of August 24, 1743 V. Rtishev, D. Korostelev, navigators K. Yushin and E. Roditchev and geodesist Gvozdev said they examined "the fiscal double and simple wine" in 60 flasks from the Yakutsk office for the expedition. A list was made of the numbered flasks. Together with the other officers, Gvozdev had various tasks. On September 2, Spanberg ordered Shelting and all officers to examine the rigging on the ships *St Joann*, *Arkhangel Mikhail* and *Nadezhda* to determine those no longer usable.[20]

In a couple of instances, Gvozdev participated in geographic and hydrographic research in the Okhotsk area. The flood of November 28, 1742 initiated a search for a better location of the shipyard and port than Okhotsk. Devier began the search. The government office at Okhotsk and the expedition united in the effort. On June 30, Shelting ordered V. Rtishev, U. Aritlander, Gvozdev and ten servitors to examine Malchikan as a location for a shipyard port and bringing ships to the mouth of the Okhota River. In his report to Spanberg, Shelting said that it was the officers' opinion that a shipyard could be built there but sailing the ships out would be impossible because of the low water level.[21] Gvozdev's report detailed the special characteristics of the river. From July 7, there were fast currents and many streams with depths of 9 feet, 6 feet and 3 feet and lower at ebb tide. There is a simple calculation that indicates that it was not economical to transport cargo in small ships (52, pp. 234-235). The next voyage was at the end of July 1743, when A.I. Chirikov ordered an investigation of the mouth of the Okhota River to determine the possibility of sailing small sea-going ships to the sea. In a report of August 5, K. Yushin and D. Korostelev described the route down the Batom River from Malchikan. The river had streams of 1, 2, 3 and 4 feet deep.[22] The navigators ruled out the possibility of sailing a ship down the river because of the swift current and shallowness of the water. A. Zybin, commander of Port Okhotsk, also decided to make a voyage to Malchikan, accompanied by Rtishev, Aritlander and Gvozdev.[23]

When Gvozdev became a member of Spanberg's group in 1741, his material circumstances improved. Salaries were usually paid three times a year in January, May and September. Accordingly, on March 7, 1743 Gvozdev received the January installment installment of his 6 rubles salary and 6 rubles bonus per month. At each of the three regular payments, he would receive 48 rubles, less one kopeck per ruble medical tax that yielded him 47 rubles, 52 kopecks. His signature appears on receipts for July 25 and in March 1744: "Geodesist Mikhail Gvozdev received his salary of 47 rubles, 52 kopecks and signed."[24] At the end of 1744, a budget was made for the following year. In this budget, Gvozdev's salary was 142 rubles, 56 kopecks. One of the geodesist's expenditures was for the purchase of groceries each month. On September 30, for example, Gvozdev purchased 56 pounds of flour, instead of the usual 12.5 pounds, 10 pounds of dried crust, 5 pounds of cereals and 2 pounds of salt. In the other months, there was no difference in the amounts he purchased. In December, Gvozdev purchased 68.5 pounds of flour, 5 pounds of cereal and 2 pounds of salt.[25] They had meat on a very limited basis. Once in a while there was an inscription about collection of money from officers "for live animals." In October, navigator M. Petrov, constable I. Bobovsky, Gvozdev and clerk A. Shubinshy together paid 4 rubles, 24.8 kopecks "for only meat."[26]

During the voyages, the officers, like everyone else on the ships, were to receive provisions in addition to their pay. In most cases in 1741 to 1743, all of those who went to sea received provisions in the form of money because of a shortage of food in Okhotsk. Based upon reports of Shelting and Rtishev, dated November 7 and December 24, 1742 on September 9, 1743 Spanberg ordered ensign S. Ivashkin on the ship *Nadezhda* to pay V. Rtishev, Gvozdev, I. Bobovsky, U. Aritlander, S. Gardebol and medical assistant I. Djagilev money for provisions not furnished during the 1741 and 1742 voyages. On another list the geodesist's name follows A. Shelting, V. Rtishev, I. Bobovsky, boatswain mate T. Gerasimov, and medical assistant P. Browner.[27] The report contained all the appropriate accounts, number of days at sea, amount of groceries, their monetary equivalents and the total. The report said that Gvozdev was at sea from September 1 to October 13, 1741 and from May 1 to September 18, 1742. The total time at sea during these two years was 6 months and 12 days. Gvozdev's salary for this time was 32 rubles, 1 kopeck. The cost of the provisions furnished is interesting:

3 puds, 24 lbs., 61.7 zolotnik crackers	3 rubles, 61.5 kopecks
67 buckets, 4 cups beer	20 rubles, 25 kopecks
two kinds of grits:16 lbs., 24 zolotnik	48.8 kopecks
32.5 lbs.	1 ruble, 5.6 kopecks
32 lbs. Beans	52.9 kopecks
19 lbs., 27.4 zolotnik butter	52 kopecks
52 cups wine	2 rubles, 8 kopecks
3 lbs., 66.85 zolotnik salt	2 kopecks

39 lbs. dry fish	1 ruble, 56 kopecks
Total	32 rubles, 1 kopeck[28]

Besides the rare materials that describe various details of everyday life and economic situations of the officers and sailors of different ranks, we can add other information describing property sold. According to sea tradition, the personal property of those who die during a voyage was auctioned to all sailors and officers who participated in the expedition. In the description of such an auction of property of deceased navigator Ivan Vereshagin in September 1743 Gvozdev purchased a small trunk for 1 ruble, 24 kopecks, white towels for 1 ruble, 20 kopecks and socks for 1 ruble, 25 kopecks. From the auction of property of deceased navigator Andrey Eselberg, the geodesist bought a red shirt for 3 rubles, 60 kopecks, braids for 36 kopecks, a pipe made from a bronze ink container for 12 kopecks and a tin dish for one ruble. In the "report to Captain Bering about the things sold" (October 1743), is a long list of items sold organized with the names of those who purchased them. Among those listed are Spanberg, Zybin, Shelting, Rtishev, Korostelev, doctor's assistant Ginter, Yushin, Roditchev, Kuzmin, Gvozdev, copy maker Dasajev, master assistant Kuznetsov and others. Those making large purchases were M. Spanberg of 160 rubles, 29 kopecks and A. Shelting of 84 rubles, 41 kopecks. Geodesist Gvozdev bought an old red jacket lining for 1 ruble, 10 kopecks in remembrance of the Captain.[29]

Gvozdev's activities satisfied the leadership of the Second Kamchatka Expedition in Okhotsk. Now the geodesist was in the Okhotsk detachment of Spanberg, which numbered 76 people including 47 who worked behind the scenes at the port and expedition area.[30]

An important event was the announcement by the Siberian headquarters of a directive from the Czar. A copy arrived at the Okhotsk office from Irkutsk on March 6, 1743. The order pardoned Gvozdev for his insubordination and other deeds of misconduct.[31] Complete clemency allowed him to request officer status from the Senate and Admiralty College. He prepared two requests but did not have an opportunity to give them to Spanberg. The Captain left for Yakutsk on March 31, 1743 arriving April 10. From there he went to Yudom where he stayed until the end of July, returning to Okhotsk in August. During this time, Zybin, the commander of Okhotsk, sent a lengthy report to Spanberg that disclosed the contents of the original documents of the voyage led by Gvozdev to America in 1732 (30, pp. 236-243). The Okhotsk office saved this precious document that includes:

1. "Announcement" of I. Skurikhin of April 8, 1741,
2. Report of Gvozdev on December 23, 1733 from Kamchatka,
3. A portion of the logbook of Gvozdev and Fedorov from August 22, 1732,
4. Report of I. Skurikhin, of April 10, 1741
5. Report of Gvozdev in 1741.

It also contained Gvozdev's activities after July 14, 1741 and the order from Irkutsk of July 6, 1742 about the investigation of the Big Land and organization of an expedition. The Vice Governor of Irkutsk asked Zybin, who in turn asked Spanberg, to make two copies of the map of the Big Land and to send them to the Siberian office.[32] With the support of the central government, the Siberian government developed instructions regarding the more important problems related to the Big Land. These instructions were the same as those given to Pavlutski. Finally, Zybin again asked Gvozdev be assigned to his command since the geodesist had already been on the islands near the Big Land and would be sent there again (30, p. 243).

When Spanberg returned to Okhotsk from Yakutsk, he started to respond to the report of Zybin and the Siberian Administration. Then on August 24, Gvozdev asked Spanberg about giving him the rank of an officer.

Gvozdev's request to the Empress, Elizabeth Petrovna, was written by copy maker Yakov Dasajevon on simple paper "because there was no other paper in Okhotsk." According to tradition, under each point of the document Gvozdev signed, "This request is signed by geodesist Mikhail Gvozdev." As in 1740, the request contained a short autobiography (since 1716) and then the experience of two geodesists in Siberia who received the rank of Sub lieutenant in 1732." And I, being in such a remote and hectic place, unlike the aforementioned geodesists, do not hold the rank of an officer," wrote Gvozdev. At the end he asked for promotion to the rank of sub lieutenant" in consideration of the difficult service and long voyages" (25, p. 42). The Captain accepted the request and immediately ordered Gvozdev to prepare a detailed report of his voyage in 1732. Spanberg asked Gvozdev "to report the details" and especially about "what you saw, accurately describing the time you were in the north, submitting your notes and map if you have them."[33]

Gvozdev prepared the fourth report (after 1732, 1733 and 1741) of the voyage to the Big Land in a period of one week and submitted it to Spanberg on September 1, 1743.[34] A.V. Efimov analyzed the important documents about the discovery of America by the Russian people including the reports of Gvozdev and witness of Skurikhin. A. Polonski was the first to discover the material in the archives in 1850. After Sokolov, Efimov noticed that Polonski combined contradictory information regarding other paricipants in the voyage without necessary analysis. Referring to Sokolov's publication, Efimov criticized the naval historian for accepting the last report of the geodesist as the most accurate. In Efimov's opinion, while the last report was more detailed than the 1741 report, it was less factual. This opinion, however, is not proven. First, Efimov noted that by 1743, Gvozdev of course could remember fewer details concerning the voyage. Secondly, he claimed that the geodesist wrote the 1743 report in fear (30, p. 163).

There is no substantiation for these arguments. First, it is not possible to seriously believe that Gvozdev wrote his report from memory. Even with a phenomenal memory, it is impossible to remember after nine to eleven years such insignificant details as the following: on August 13, "between four and five in the afternoon, the weather was calm and we dropped anchor." On August 15, "at the beginning of the 11th hour we got the right wind. "On August 17, "after midnight, around 7 o'clock, we visited the island." "After midnight, between 2 and 3, everything was still." Skurikhin, who gave his report in 1741 at almost the same time as Gvozdev, could only remember the month they returned and the names of many of his comrades on the voyage. Gvozdev, of course, did not write his report from memory. He based it upon his own notes and documents saved in the archives in Okhotsk. The texts of the documents differ, but the general content is the same. Using these notes, Efimov tried to prove that Gvozdev had either one of his earlier texts or the ship's log before him while completing his 1743 report. It is unlikely that he had the ship's log before him because of the categorical statements he had made in the past denying one ever existed. He said that without this log, he could not relate "all the circumstances in this report." Geographical historian, V.I. Grekov, noticed that the information written by the geodesist in 1741 and in 1743 was very similar and contained only minor differences. In his opinion, the editing of 1743 was more precise than that of 1741 (23, p. 347).

It seems an exaggeration of the truth that Gvozdev wrote the report out of a "sense of fear" as Efimov indicated. He believed that the conflict continued with Devier for more that two years based on Gvozdev's refusal to prepare a map according to the ship's log. By September 1743, the geodesist must have been free of any fear that may have existed at one time because six months before writing the report he had received clemency from the Siberian office of the government.

Gvozdev understood very well that he hadn't answered some of the questions he was asked in Irkutsk and Okhotsk before the Second Kamchatka Expedition. Trying to excuse himself, the geodesist again recounted the history as it related to Fedorov and cited "discrepancies" in the ship's log as the reason "it was impossible to prepare a map." He had explained the omissions in the report of compass readings and latitudes of the discovered islands through references in the report to G.G. Skorniakov-Pisarev in 1733. It is important to emphasize one characteristic in the report of 1743 that has previously gone unnoticed. The report of 1743 differs from the report of 1741 in that the former does not contain any information about Petrov, who had given Gvozdev so much trouble.

On September 7, after reading Zybin's information and Gvozdev's report, Spanberg gave his written opinion to the Okhotsk office.[35] During September through November, 1743 in a couple of documents he sent to Okhotsk, Irkutsk and St. Petersburg, Spanberg repeated the rationale to organize a new expedition to the Big Land. First, Spanberg understood and accepted Gvozdev's explanation that it was impossible to make a map of the 1732 voyage because of the absence of resources. He then ordered an exact copy of the geodesist's report for the Okhotsk office.

Bering's successor was very much opposed to the transfer of Gvozdev to Zybin. He thought it was impossible for him to allow the transfer of the only geodesist he had who could make copies of maps. Accordingly, he would not send Gvozdev to Zybin. At this time he thought Gvozdev was indispensable, "because the expedition needs a geodesist to make copies of voyage maps when necessary, and there is only one such geodesist."[36]

Especially interesting and important were two suggestions regarding the immediate preparations for a new voyage and the possibility of making a map of the 1732 voyage. Spanberg thought that if preparations for the planned expedition were to go slowly because of the Okhotsk office, it would be possible to make the Second Kamchatka Expedition responsible for the realization of the new voyage to the Big Land. Then we find an unusual bit of information received from an unexpected source - the diary of Fedorov that was obtained from assayer S. Gardebol. Spanberg decided to give this, "Fedorov's own journal, written for his own needs [separate from the journal written together with Gvozdev]" to navigators K. Yushin and E. Roditchev and to Gvozdev for expertise and taking an "extract" from it and making a map according to "all the described places in this journal discovered during the voyage."[37] There was a real possibility of making a correct map of the 1732 voyage to the shores of Alaska.

The only puzzle is how Fedorov's private diary came into the hands of S. Gardebol and from him to Spanberg. It is impossible to answer these questions because we have neither a copy of the diary nor an "extract" of it. Research of the biography of Gardebol reveals nothing. Presumably, the diary came to Spanberg sometime between 1738 and 1742.

On September 9, 1743 Spanberg gave Fedorov's diary to Yushin, Roditchev and Gvozdev, directing them to make a map of the voyage from the Kamchatka River to the shores of Chukotka, to the islands and to the Big Land. After three days, Yushin asked for the extract and two sheets of Alexandria paper or half-Alexandria paper in order to accomplish the task. On September 13, Spanberg ordered "for the making of an extract of the voyage of 1732" some writing paper be given to Gvozdev and because of the lack of special map paper he was given six sheets of "simple, good paper."[38]

On October 8, Yushin reported to Spanberg giving him a map "from the Kamchatka River to Chukotka and farther to the islands, and the Big Land," Fedorov's diary and the extract. The navigator said that they had difficulty making the map because the diary "was written very haphazardly and it is very difficult to make an exact map according to it."[39] On October 13, when Zybin again emphasized the importance of information about the unexplored islands and regions, Spanberg responded with pride that "just now some information has been found" and on the appropriate occasion it would be sent "where it has to be sent."[40]

This is the history of the creation of the new map of the 1732 voyage that was made by navigators Yushin and Roditchev and geodesist Gvozdev. But neither the original of the map nor copies can be located, only its general appearance in the form of a map well known in historic, geographic literature as the "map of Spanberg of 1743" (30, p. 168). Divin criticized Efimov for his identification (25, p. 38). However, the documents preclude distinguishing the original from the copies.

In the beginning of November 1743, a situation arose that made it possible to send the reports from Okhotsk to Irkutsk and St. Petersburg. On September 7, Spanberg prepared and sent a report (No. 574) to the Admiralty College and an account (No. 606) to the Irkutsk Province office with the explanation of the circumstances of the creation of the map of the 1732 voyage from Fedorov's diary. The contents of these two documents are not very different from each other or Spanberg's resource document.[41] However a small difference can mean a lot. In addition Spanberg sent the map "made by" Yushin, Roditchev and Gvozdev to the Irkutsk office. Because "all the sea voyage descriptions belonged to the

Admiralty College," Spanberg included with his report copies of Zybin's account dated April 20, Gvozdev's report of September 1, and "the same map" which was sent to Irkutsk with the "addition of the voyage of Captain-Commander Bering, who happened to be in this same area in 1729 and 1741."[42] The notes about sending documents from Okhotsk prove that "the map of Gvozdev's voyage" was sent to Irkutsk and "the maps of the voyages of Gvozdev and Captain-Commander Bering" were sent to St. Petersburg.[43] So the documents inform us that in October 1743, in Okhotsk, the geodesist together with navigators Yushin and Roditchev made a map on simple writing paper of the voyage of the ship *St. Gabriel* from the delta of the Kamchatka River to the islands and the Big Land. The second map signed by Spanberg was made from a compilation and reflected not only the voyage of 1732 but also Bering's voyages of 1729 and 1741.

On November 4, Gvozdev heard about this and requested Spanberg to submit the two petitions he had prepared in August to the Senate and Admiralty College "with your description of my good character."[44] Spanberg supported the geodesist and wrote the following to the Senate and Admiralty College: "He, Gvozdev, has been here in these difficult circumstances for a long time at his rank of geodesist. All his colleagues, having served fewer years, are now lieutenants and captains. That is why I ask that he, Gvozdev, for this long service in these arduous places, be promoted in his rank and income according to the number of years he has served and to his colleagues' careers. He deserves it very much because of his industrious and intelligent work during the sea voyages and land trips."[45] This recommendation probably wasn't formal. Despite having many unpleasant traits, Spanberg was an excellent organizer, a smart and experienced naval officer and an excellent mapmaker. The high recommendation of Gvozdev's service is significant because high recommendations from Spanberg were rare and unusual and only given in cases when they were deserved.

Footnotes to Chapter 5

1. TsGADA, f. 192, Irkutsk Province., No. 104. The other copy of the plan is published by F. Golder and S.V. Zmanenski (23, pp. 87-8) from the fond of Bering (TsGAVMF, f. 216, Case 24, p. 1019).
2. TsGADA, f. 248, Bk 1361, pp. 43-44; Bk 1367, pp. 1121-2; f. 24 Bk 14, Part 19, the other side of p. 323.
3. TsGAVMF, f. 216, Case 56, p. 1074.
4. Ibid., pp. 1074-1077 and the other side.
5. TsGADA, f. 248, Des. 113, Bk 1552, p. 1 and other side of 2
6. TsGAVMF, f. 216, Case 99, pp. 355 and other side of 358.
7. TsGADA, f. 214, Description 1, Part 7, p. 5122, pp.1-2, f. 248, Description 113, Bk 1552, p. 2.
8. Ibid., f. 214, Des. 1, Part 7, Case 5122, pp. 1 and other side.
9. Ibid., f. 24, Bk 14, Part 22, the other side of p. 174.
10. Ibid., f. 248, Description 113, Bk 1552, other side of p. 2.
11. Ibid., pp. 7-9.
12. TsGAVMF, f. 913, Des. 1, Case 38, pp 440-460 and other side.
13. Ibid., p. 460.
14. TsGADA, f. 248, Bk 1326, the other side of p. 21.
15. Ibid., pp. 22-23.
16. Ibid.
17. TsGAVMF, f. 913, Des. 1, Case 38, pp. 495, other side of 521.
18. Ibid., p. 521.
19. Ibid., f. 216, Case 53, the other side of p. 730.
20. TsGAVMF, f. 216, Case 99, p. 233, 348-350; Case 53, the other side of p. 521, Case 56, the other side of p. 807.
21. Ibid., Case 56, p. 59.
22. Ibid., p. 58.
23. Ibid., pp. 58 and the other side, pp. 67 and the other side.
24. Ibid., Case 99, pp. 68, 102, and the other side of 259; Case 56, and other side of p. 259, Case 101, p. 187.
25. Ibid., Case 99, pp. 254, 298, 301, 455, 498; Case 101, pp.126,185,244
26. Ibid., Case 99, p. 336.
27. Ibid., Case 56, pp. 577-578, 581 and the other side.

28 Ibid., Case 99, Page 270; Case 56, p 589.
29 Ibid., Cases 56, pp. 549 other side of 461, other side of p. 458.
30 Ibid., pp. 5-7.
31 Ibid., Case 53, the other side of p. 730-1.
32 Ibid., Page 731.
33 Ibid., Case 99, the other side of p. 238.
34 Ibid., Case 53, pp. 733-738 and the other side.
35 Ibid., Case 53, pp. 730-742; Case 99, other side of pp. 259-261.
36 Ibid., Case 53, p. 741.
37 Ibid.
38 Ibid., Case 99, p. 266; Case 56, pp. 112 and the other side.
39 Ibid., Case 53, the other side of pp. 743-744.
40 Ibid., The other side of p. 745, Case 99, p. 315 and other side.
41 Ibid., f. 216, Case 99, the other side of pp. 355-358, published without exact date (52, pp. 246-249).
42 Ibid., p. 358.
43 Ibid., Case 53, the other side of p. 742.
44 Ibid., Case 56, p. 968.
45 Ibid., Case 99, p. 347, the other side of pp. 366-367.

Chapter 6
Retirement

On September 24, 1743 the Czar ordered all activities of the Second Kamchatka Expedition to cease. He ordered Spanberg and Chirikov "not to make any more voyages unless commanded to do so because of the food shortages in Irkutsk and Eniseisk Provinces." The members of the expedition were to live as a group in Siberian cities. On September 26, the Admiralty College and the Siberian government selected the cities of residence and made arrangement for payment of salaries and other organizational matters. By the summer of 1744, news about the disbanding of the expedition reached Okhotsk. At that time, there were 161 people in the expedition. Only 48 of them were actively working, includeing 12 describing the Penzhin Sea, 6 in Okhotsk, 2 in Kamchatka and so forth. Gvozdev's name appears on all lists of the expedition in 1744.[1] On July 10, Spanberg left the sailors and clerks in his command under V.R. Rtishev and departed for Eniseisk with a portion of his detachment. On July 13, A. Shelting, who was responsible for a special ship, received an order from Spanberg "to gather the rest of the dispersed group and bring them to Eniseisk immediately."[2] Shelting's list of 34 subordinates begins with geodesist Gvozdev. On August 21, Spanberg told the provincial office in Eniseisk that he was traveling by land with some members of the expedition to live in the town of Tomsk. He requested 39 wagons with drivers, including 7 for himself and one each for A. Kuzmin, P. Grigorjev, S. Ivashkin, E. Ginter, M. Gvozdev and others.[3] On August 29, he sent the first group of 20 people and 13 wagons headed by Grigorjev and Gvozdev. The next day he sent a letter to the military commander in Tomsk requesting that he prepare for their arrival "good apartments for them in one place together near the offices."[4]

On September 11, Captain Spanberg arrived in Tomsk. The next day he ordered the head of the Tomsk military command to build stoves in buildings occupied by members of his expedition where there were none. This started a new period in Gvozdev's life in Tomsk where he lived for 10 years and 3 months. There is not much information about him during this period except his presence in Tomsk from 1744 to 1755. Gvozdev spent all these years without leaving Tomsk. The commanders changed, the number of people in the group changed according to the tasks assigned, but Gvozdev's name appeared on all the rosters and monthly reports sent to the Admiralty College and Siberian Governor.[5]

In Tomsk, they began a slow, regular garrison duty with little excitement in contrast to the turbulent days of service in Kamchatka and Okhotsk. Spanberg put Gvozdev in charge of the expedition's finances. On September 20, Gvozdev and Lt. S. Ivashkin were ordered to audit the accounts and record receipts of funds used to move the 39 wagons and personnel from Eniseisk to Tomsk, a distance of 635 versts. In charge of the funds were assistant compass maker S. Kuznetsov and Academy students A. Fenev and P. Shananjkin. On October 24, Gvozdev received another task. With S. Kuznetsov and E. Roditchev, Gvozdev was to audit the book of debits and credit maintained by F. Voronov and to prepare an accounting report.[6] On November 6, Spanberg ordered another report: "Tomorrow, assistant Voronov will audit expenditures from the Czar's budget. This audit will be conducted in the presence of Roditchev and geodesist Gvozdev. After the audit is completed, give an accounting to me." Another order was given on November 9: "All senior and junior officers will meet tomorrow including ensigns Grigorjev and Ivashkin, navigator Roditchev and geodesist Gvozdev."[7] As the result of the audit, Gvozdev, Roditchev and Kuznetsov discovered an ominous shortage of three rubles, 11 11/120 kopecks. Spanberg took this sum from F. Voronov. A. Fenev, the Japanese language student, then became responsible for the Czar's budget[8]

In 1745, Gvozdev became eligible for a yearlong vacation based on his 28 years of service. However, he did not use it. As an orphaned son of a soldier, he probably had no one to visit and Siberia and his service had been home to him for a long time. His health deteriorated with many years of food shortages, difficult living conditions and scurvy. His eyesight was weak and growing worse. In his 1758 petition concerning retirement, Gvozdev wrote that in Tomsk, "My body is weak, I cannot see very well and have old scurvy and kidney disease from long years of service and distant sea voyages" (52, p. 158). In especially bad cases, he received medical attention. Dr. E. Ginter reported that Gvozdev was in the hospital from February 9 to 15 and from April 22 to May 3, 1745.[9]

Before leaving Eniseisk for St. Petersburg, A.I. Chirikov gave S. Waxell all documents of the Second Kamchatka Expedition with which he had been entrusted after Bering's death. Here it is mentioned that, "The first map made by the order of Fleet Captain Chirikov of the western shore of the Penzhin Sea from Okhotsk was made after Rtishev and Shelting were sent there. The region east of the eastern corner of the Chukotka Peninsula between 65° and 68° was made from the map drafted by geodesist Gvozdev."[10] Undoubtedly, "the map of geodesist Gvozdev" is the map of the 1732 voyage drawn in 1743 by Gvozdev, Yushin, and Roditchev using the private, "unofficial" diary of Fedorov.

For more than three years Gvozdev waited for a decision regarding his request for a promotion from the Senate and Admiralty College. So when, by his account, the time for an answer had elapsed, he decided to continue his petitions. On February 12, 1747 an entry was made in the journal of the commander of the Tomsk group about the new petition of the geodesist. Gvozdev asked "about inquiring of the Admiralty College regarding his promotion and bonus, and those of his colleagues, Sub lieutenant Skobeltsyn and four others."[11] In all probability, this petition was not sent because on July 5, 1747 a related entry appeared in the journal: "The petition of geodesist Mikhail Gvozdev asking about the sending of his request to the Admiralty College about the promotion of his rank and bonus."[12]

For a long time Gvozdev was responsible for various tasks relating to making copies of map materials of the Second Kamchatka Expedition that Waxell had brought from Eniseisk. Gvozdev did not put away his geodesist instruments. The list of equipment in Gvozdev's possession included, "one feodalit, one bronze quadrant without tripod and a drafting box with compasses."[13] Gradually, Gvozdev became responsible for receipts and disbursements. Gvozdev reported to Waxell that he had debits of 8,374 rubles, 72 7/15 kopecks in the Czar's budget for 1747-1748 and credits of 6,596 rubles, 2 2/3 kopecks for the same period. Waxell wrote to St. Petersburg that "much of the leftover funds were under the control of geodesist Gvozdev and skipper Dmitri Dorostelev."[14] By August 20, 1748 the geodesist had a surplus of "1,778 rubles, 4/5kopecks."[15]

In his 1758 autobiography, Gvozdev wrote that he had the rank of geodesist until 1732, "and in that year according to an order of the Senate he received a promotion to the rank of sub lieutenant for his long service in Siberia" (52, p. 157). That announcement seems to contradict the other documents after 1732 where Gvozdev was called (he called himself and signed) a geodesist. All rosters of the Tomsk group through 1749 identify him as a geodesist. In the roster from December 1, 1749 Gvozdev was still referred to as a geodesist. However, beginning with the roster of May 1, 1750 he was a sub lieutenant. Unfortunately, the rosters for January through April 1750 have not been found, making it impossible to determine the precise date of his promotion. We can only say that the order from the Senate concerning Gvozdev was received in Tomsk between January and May 1750. After that time, "The Sub lieutenant of cartography with his servant and for food" began receiving an annual salary of 94 rubles, 3 kopecks instead of 71 rubles, 28 kopecks.[16] At last, he realized his burning desire and won justice. Having been entitled to the rank of sub lieutenant of cartography and thinking of himself in those terms since 1732, Gvozdev finally received the promotion in Tomsk in 1750. I found indirect but certain information about this. According to the 1758 report, the Senate order of 1749 gave Gvozdev, in addition to his new rank, the right to another promotion after 1732 before that of his colleagues. In addition, Gvozdev could have requested the difference in salaries between a geodesist and an sub lieutenant from 1732 which would have amounted to more than 600 rubles.

In the middle of the eighteenth century, the task of industrial development in Eastern Siberia required a lot of hydrographic, geographic and cartographic research. All members of the Second Kamchatka Expedition, including those in Tomsk, transferred to F.I. Somoinov, Governor of Siberia and leader of the Nerchinsk

Expedition, to work in Irkutsk Province and the regions of Nerchinsk and Selenginsk (21, pp. 105-130). On August 18, 1754 the Senate ordered additional descriptions of rivers and lands "to find land useful for agriculture," to make provisions for transportation between Irkutsk and Nerchinsk "and to describe all the land in the Irkutsk District. "Because of the shortage of cartographers who specialized in large territories, on December 9, 1754 Gvozdev and Lt. N. Chikin were sent from Tomsk to Irkutsk. After January 1755, the names of both geodesists appeared on the roster of those "serving under Major General and Vice Governor Wolf in Irkutsk."[17]

On April 5, 1755 the Irkutsk office instructed Gvozdev to "determine which lands in the Irkutsk District were suitable for agricultural purposes and to prepare drawings." With much difficulty, he described the "various stations and Uritski and Kudinski villages along the Moscow Road" during the next three years. But as he explained in his 1758 biography, he couldn't complete the work because of poor eyesight and failing health (53, p. 159). M. Outsin continued the work started by Gvozdev in the Irkutsk region in 1755-1757. V. Vsinov then finished the task. On December 4, 1768 Vice Governor A. Brill noted in his report to the Senate that despite the fact that Gvozdev was "determined to do the land descriptions," he was sent to Tobolsk because of his age without finishing the job. Vsinov was sent to Irkutsk in his place. In 1760, Vsinov reported that Gvozdev's journal "contained good descriptions and that without them, he couldn't work on the plans" (57, pp. 32-33). The replacement by the younger geodesist was probably a good idea since Gvozdev's last years of work had been very hard on him. Gvozdev was aware of his declining productivity. He wrote "because of my illness, I probably would not be able to work anymore as a geodesist" (52, p. 159).

In June 1758, Gvozdev asked Irkutsk's Cossack, I. Petukov, to write to the Empress, Elizabeth Petrovna, "to allow him to retire" because of his illness and long service. The request was sent from the Irkutsk Administration to Tobolsk. Siberian Governor F.I. Somoinov, who knew the geodesist very well, supported him. In his report of December 23, 1758 Soimonov requested Gvozdev's retirement and that he be allowed to do the work that "he was still able to do" (25, p. 47). The decree of the Senate of March 29, 1759 agreed to the request for his retirement, and "his future employment" was left to F.I. Somoinov.[18]

There is no information available concerning the death of Mikhail Spiridonovich Gvozdev. Of his 43 years of government service, he lived and worked for 32 years in Siberia and the Northeast region of Russia.

Footnotes for Chapter 6

1. TsGAVMF, f. 216, Case 101, pp. 55, 278, and 283. From the fall of 1743 to March 4, 1744, Gvozdev was in a hospital on the Island Bulgin in the mouth of the Okhota River.
2. Ibid., pp. 274 and the other side of 275.
3. Ibid., the other side of p. 354.
4. Ibid., pp. 363-364, 374 and the other side of 375.
5. Ibid., Cases 63, 66, 73 and 101 contains data from 1744-1757.
6. Ibid., Case 161, pp. 382 and 404.
7. Ibid., Case 102, the other side of p. 7.
8. Ibid., pp. 168 and the other side of 174.
9. Ibid., pp. 11 through the other side of 13 and 25-26.
10. Ibid., Case 105, other side of p. 70; Case 66, other side of 205.
11. Ibid., Case 66, the other side of p. 313.
12. Ibid., p. 316.
13. Ibid., p. 59 and 100.
14. Ibid., pp. 57 and 96 and the other side of each.
15. Ibid., pp. 399-410.
16. Ibid., Case 72, pp. 146 and 196
17. Ibid., Case 73, pp, 227 and 229; TsGADA, f. 248, Part 113, Bk 485a, p. 396.
18. TsGADA, f. 248, Part 113, Bk 485a, p. 401.

Chapter 7
The Pioneers of Alaska from the Pacific Ocean Side

The discovery of Alaska from the Pacific Ocean side was a natural, continuing process of joining Siberia to Russia. The history of the Kamchatka Expedition with the leadership of Gvozdev is a bright example of people's movement to the east and the government's colonial policy in Siberia during the 17th and 18th centuries that resulted in the geographic discoveries in Northeast Asia and the Arctic and Pacific Oceans.

D.I. Pavlutski and the Okhotsk Administration did not sufficiently appreciate the information reported about the voyage of 1732. The journal from the *St. Gabriel* and the reports of Gvozdev and Skurikhin were probably lost. The Admiralty College was unsuccessful in attempts to obtain these documents. In the 1760's, Siberian Governor F.I. Somoinov, on one of his maps about the discovery of North America in 1732 by Gvozdev mentioned, "he couldn't find the original journal."[1] In the nineteenth century, historians studying the archives of the Naval Department said, "they couldn't find the journal which they thought must have been very interesting" (12, p. 10).

Some researchers have tried to discount the value of this important historic, geographic event by emphasizing an erroneous concept of the occasion of the discovery. They thought that Gvozdev couldn't have known that he had discovered the American continent from the side of the Pacific Ocean. Thinking in this manner, F. Golder considered the discovery of America by the Russians as an every day event. In his interpretation, Gvozdev did not understand that he had seen the coast of the continent. Referring to G.F. Müller's story, F. Golder comes up with inaccurate information about Pavlutski's order to Gvozdev to bring the surplus from Bering's expedition to the Chukotka Peninsula. In addition, he said that Gvozdev couldn't find the commander of the Anadyr detachment and turned back and accidentally arrived on the American coast (75, p. 151). Academician Müller, the famous historian, a person who knew the geography and cartography of Siberia very well, did not have sufficient information about the

Gvozdev voyage. Only by the 1850's could he have received some unofficial information about the voyage of 1732. He was correct in his thinking that the Big Land was the coast of North America and that "this land has significant value since the Chukchi obtain animals from there that live only in warmer regions" (30, p. 226). But the Academy of Sciences, the Siberian government and other government departments were unaware of the results of the voyage of 1732 for several years.

In the first half of the eighteenth century, the Admiralty College, Siberian Administration and later the Senate were very interested in Gvozdev's voyage mentioning it three times: In 1738, because it was necessary to verify the report of sailor L. Petrov concerning Shestakov's trip of 1730; in 1741, because of the organizing of a new expedition to the Chukotka Peninsula under the leadership of Pavlutski; and in 1743, because of the discussions of a new expedition to the Big Land. But before then, officially reported materials about visiting the Alaskan coast by Russian sailors came to St. Petersburg and Tobolsk. This exact information also went to France. Among the documents of French scientist J.N. Delisle, who served in the St. Petersburg Academy of Sciences, was this notice: "Voyages and discoveries made by Russians in East (Pacific) Ocean during 1731 and 1732 between Bering's two voyages" and "Map, illustrating Gvozdev's voyage" (75, pp. 154, 299-301). It is known that a French newspaper reporter made these documents from an interview with F.I. Somoinov on March 1, 1738 (21, p. 89).

Somoinov didn't mention any names of the participants in the expedition or any information about the discoveries in the interview. Even so, it was the earliest foreign publication about the first successful voyage to America from the Pacific Ocean side. There is no exact information about why the original materials reported didn't come to St. Petersburg in 1738. The basis of interview were the incomplete reports of L. Petrov and information received from Bering and Spanberg. These notes of events can be compared with the real ones and the expeditions of Shestakov-Pavlutski and the meeting of Bering with ten sailors; voyages in the Okhotsk Sea to Shantar Island and the Kurile Islands - the voyages of Ivan and Vasil Shestakov. There are no clear explanations about the German Navigator (probably Gens) who arrived in Tobolsk with the ship's journal and the appearance of the mariner in the capital. The rest of the information about the ship left by Bering, discovery of the islands and land, meetings with native people in a small boat, sailing near the coast in a southerly direction, the storm and coming back to Kamchatka - all refer to Gvozdev's voyage. The map illustrated in the interview is one of the attempts to interpret the discovery of Alaska during the voyage of 1732 on the route: the mouth of the Kamchatka River, the Chukotka Peninsula, the islands in Bering Strait, the Big Land and the mouth of Kamchatka River. This map shows not only Gvozdev's voyage, but also a number of different routes of the Shestakov-Pavlutski Expedition in 1730-1732. The illustration of only one Diomede Island instead of two (King Island is also missing) is among of the errors in the description of the Gvozdev discoveries.

It is not possible to retrieve the exact route of the *St. Gabriel* to the Seward Peninsula since the reports with the navigation information are still missing. Neither Gvozdev nor Fedorov completed a map of the voyage from the mouth of the Kamchatka River to the Big Land. It is known only that in October, 1743 by Spanberg's order, Gvozdev with navigators Yushin and Roditchev prepared the map of the voyage in 1732 to the Big Land which was based upon information from Fedorov's "private" diary. This has also been lost but Spanberg's map of 1743 presents the same results (5, No. 69). Spanberg's map of 1743 was quite exact about reflecting the results of the voyage of 1732 and a portion of the voyages of Bering and Chirikov. There are still five copies of this map which differ from each other only in some small details.[2] "The map from Okhotsk to Lopatka and to Chukotka Peninsula." has a name which mentions the main sources: "previous" description in 1728 of the First Kamchatka Expedition - Okhotsk region – Cape Lopatka - part of Chukotka Peninsula; according to Fedorov's journal islands and the land "near" Chukotka Peninsula; the map of 1741 made on the ship *St. Peter* describing the territory to the east of Cape Lopatka "to the distance of 35°00'."

This map presented Okhotsk reflects the mouths of the Okhota and Kuhtui rivers, some of the northern most Kurile Islands, the west coast of the Bering Sea from Cape Lopatka to Bering Strait, the Diomede Islands, the Big Land, and St. Lawrence, Commander and Aleutian Islands. On the map there are depth marks along the gulf of Anadyr and Bering Strait.

Spanberg's map, as does Gvozdev's reports in 1741 and 1743, proves the discovery of North America by Gvozdev during his voyage on the *St. Gabriel* in 1732. The map scale could not present many details (60-62 miles per inch). Probably the most interesting features of this map are how the three islands in Bering Strait, the Asian and American coasts and part of the route of the *St. Gabriel* are presented. It does not show the entire route of the *St. Gabriel* and the portion presented is shown in a dotted line in the general direction of the voyage and goes along the Big Land to the southeast.

There is a note on the map, "Here was geodesist Gvozdev in 1732" near the northwest part of the Big Land. After that, the name of the Russian geodesist is forever associated with his discovery of Alaska. His discovery is an important and valuable contribution in the eighteenth century world of geographic science and was recorded on maps in Russian, French, English and German languages. In spite of the fact that the information and map of Gvozdev's voyage were secret, already by 1746 on the map of the world ("Mappa Monde...") on the east coast of Bering Strait appeared the note "detecta a Gwozdew 1730." "The land discovered by Gvozdev" is on the map of 1754-1758 by G.F. Müller,[3] being published in French and English (42, p. 103). Publishing in 1752, "Carte generale desdescouvertes de l'Amiral de Fonte et autres navigateurs Espagnole, Anglais et Russes, pour la recherche du Passage a la Mer du Sud," J.N. Delisle tried to combine (without much success) many events with fanciful information about contributions of European discoveries in the Pacific Ocean and American continent. This map contains many errors.

Without mentioning Gvozdev's name, Delisle presents the route of the *St. Gabriel* in 1731 (correctly 1732) from the mouth of the Kamchatka River to the Chukotka Peninsula, which is the same as the route of Bering's voyage in 1728, then the direction of the ship along the Alaska coast and "the return of the Russians to Kamchatka in 1732." The map was made according to an interview with Somoinov written by Delisle in 1738. The scheme of this map and the one of 1752 do not have accurate information as their source since the basis of both was approximate information.

"Map from Okhotsk to Lopatka and to Chukotka Peninsula..." October 1743
The main source - map of 1743 by K. Yushin, E. Rodichev and M.S. Gvozdev.

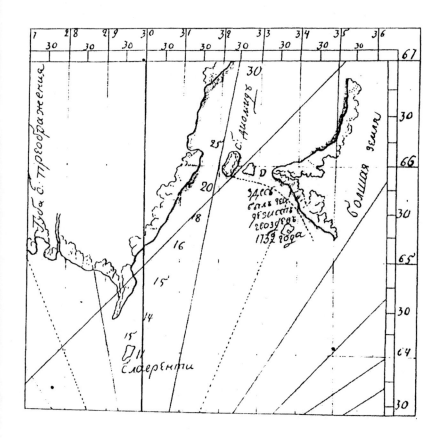

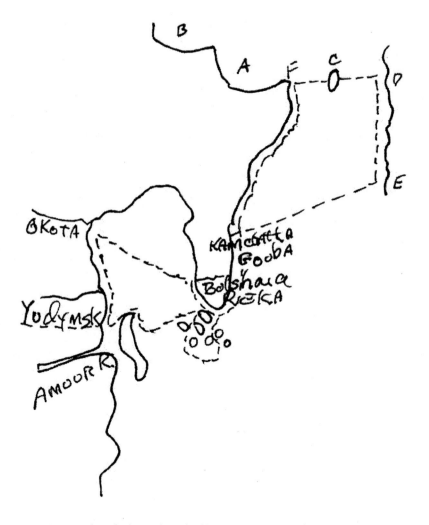

Copy of Delisle's Sketch Illustrating Gvozdev's Voyage

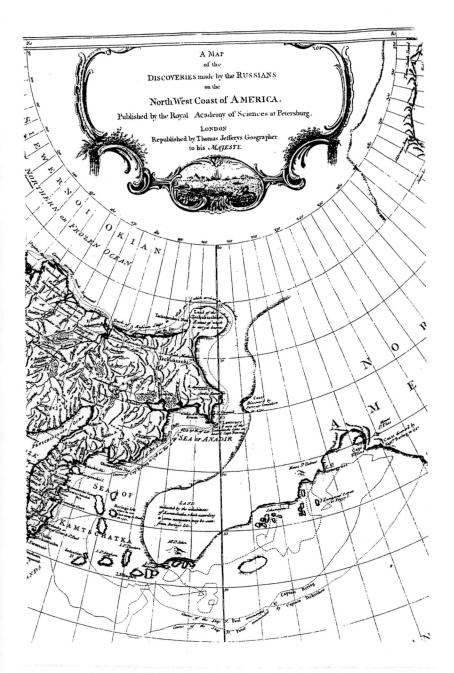

Portion of Truskott-Müller Map of 1754-1758

There is some tradition (7, p. 142) of presenting the Chukotka Peninsula, Bering Strait and the Big Land (Alaska) in Russian map science (64, p. 111). One of the different types was a Russian map completed according to the old tradition and not according to var-ious mathematical and geographic rules.

Among the maps of that period are "Kamchadalii" of 1732; A.F. Shestakov's map of 1724, I.K. Kirilov's map of 1724; I. Lvor's (not earlier than 1728) map; the map of Ya. I. Lindernau of 1742; T. Perevalov's maps of 1744, 1746 and 1756; and N. Daurkin's map of 1765 (5, No. 55, 58, 68, 122-124,127-129).

Examples of the scientific approach to mapmaking were Spanberg's map of 1743, or to be more exact, the map of Gvozdev, Yushin, and Roditchev of the same year that is still missing and the final maps of the Second Kamchatka Expedition. These presentations are very similar being based upon similar information. Referring to results in 1746 of the accomplishments of the expeditions between 1725 and 1743 in the northeast, A.J. Chirikov said, "A large part of the world was found, a lot of islands and land previously unknown," and that significant scientific results came from the work of Gvozdev. The Marine Academy Commission reviewed all the geographic materials and maps and prepared the "General Map of the Russian Empire of the North and East Siberian Coast." They listed the different resources used. Among the following are those which refer to Gvozdev's work: from Okhotsk, the west coast of the Penzhin Sea copied from the maps sent by Rtishev and Shelting; and the land adjacent Chukotka to the east, situated from 65° to 68° north latitude from Gvozdev's map (35, pp. 795-7; 29, p. 191, 26, p. 240). Similar information is in the first variant of the General Map that was made by A.J. Chirikov's order and sent to S. Waxell in St. Petersburg before his leaving for Eniseisk in the summer of 1745. This is in addition to the Marine Academy Map of 1746. In this way, the portion of the General Map of the west coast from the Penzhin Sea and Okhotsk to the mouth of the Uda River and Shantar Islands was from the research of Rtishev and Shelting with Gvozdev's participation.

A comparison of the final 1746 map with Spanberg's map of 1743 (completed by Gvozdev, Yushin and Roditchev) presented identical information on the coast from the mouth of the Kamchatka River to Chukotka, the islands in the strait and the coast of Alaska. But there is a little confusion. On the Marine Academy Map, the Diomede Islands are shown twice, as Bering's discovery and as Gvozdev's. In 1728, Bering discovered only one island in the strait, the west one, and called it St. Diomede. Upon seeing three islands on Spanberg-Gvozdev's map (now Ratmanov, Kruzenshtern and Fairway Islands), scientists of the Academy put these islands on their map on the latitude of Cape Prince of Wales, north of "St. Diomede" Island. At the same time, Chukotka and the Alaskan coast, copied from the Spanberg-Gvozdev map of 1743 to the Marine Academy's General Map of 1746, were more precise than on any other sources. It was a pity that scientific results of Gvozdev's expedition and the Second Kamchatka Expedition are not reflected in the Atlas of the Academy of Sciences in 1745 nor the map of the Geographic Department in 1754-1758 (22, p. 54).

Müller's maps of 1754-1758 and Truskott-Müller's maps of 1773 became wide spread over all Russia and Europe. They presented an incorrect picture of Northeast Asia and the Arctic and Pacific Oceans. Errors regarding the Chukotka Peninsula and many others are explained by the Academy of Sciences and academician Müller's inability to obtain the most recent information available from the Admiralty College.

It would be an error to assume that the discovery of the American coast and the many islands by Gvozdev became known only after the publishing of articles about the expedition of 1732. The 1741-1743 reports of the geodesist sent to the Admiralty College and also his map became the subject of careful research by future participants of various expeditions to Northeast Asia and the North and Pacific Oceans. For more than fifty years, this material was the only source of exact information about this unknown region. Working on a development project of Northeast Asia (June, 1746), A.J. Chirikov became interested in the fact that the expedition found a lot of different kinds of animals in the region of Bering Strait and "many other places" (26, p. 219). He wrote, "It is very important that there are many animals such as beaver, walrus, whales and many others in the land which Gvozdev saw, situated not fewer than 1700 versts from the Anadyr ostrog, by way of river and sea; more than 1100 versts from the mouth of the Anadyr River; and more than 600 versts by sea alone. There are a lot of animals near Chukotka Peninsula."[4] Chirikov came up with a new task: a new expedition to the American coast for better information. "The time will come to send another expedition to that land which Gvozdev saw at 64° north latitude, though not very soon," he wrote, "for the land, quite distant from Kamchatka and with a cold climate, has a lot of sables and other animals there."[5]

There are many other examples of intensive use of Gvozdev's materials. There are still many copies and notes from reports of the geodesist with many different marks made by people who were interested in the material and referred to it in 1850-1870. Among them, "a notice from Gvozdev's report given to Captain Spanberg," which was prepared for Anadyr's commander F. Plenisner, the notice for the Krenitsin-Levashov Expedition, about Gvozdev's report to Spanberg, etc.[6] On the backside of P. Fondezin's copy of Spanberg's map there are two notes. The first note is inconsistent with the document: "No. 1 map of Gvozdev's voyage in 1732 from Okhotsk to Kamchatka." The second note: "Copies of maps given to expedition, originals were left in Petropavlovsk" (46, No. 69). The second note says that the Admiralty College used the map materials in preparation for other expeditions. They made a lot of copies of the original map for the Kamchatka Expedition of 1754. The Admiralty College made many copies of the map for Siberian Governor V.A. Myatlov and for the commander of the Nerchinsk Expedition, F.I. Soimonov. These facts are of significant scientific value. Among the materials used that are worth mentioning are: "the list of journals of officers who participated in the Kamchatka voyage; Ship's Journal No. 15 prepared by midshipman Rtishev and geodesist Gvozdev in 1742; the list of Kamchatka Expedition's plans and maps, from which copies were made," and containing Map No. 5, the "Map of the route by the way of the Penzhin Sea from Bolsheretsk to the Kurile Islands, made by Rtishev and Gvozdev,"[7] and map No. 42, "two plans of convenient locations for residence near Okotsk" prepared by Gvozdev (with some marks: "one map is left in the College and the other sent to Myatlov without copying"[8]). There is a copy of Spanberg's map of 1743, made by F. Trapeznikov in 1754 (46, No. 70).

The N.P. Shalaurova's Expedition used Spanberg's map of 1743 and Gvozdev's description of the voyage. The coast of the Big Land and islands in the strait on the map of Northeast Siberia are identical to the map of 1743 (72, p. 322). Philip Vertlugov in 1767, the writer on the expedition, made the map. Besides that, Vertlugov probably marked the voyage of 1732 on it. On the map one can see "the route of the ship *Gabriel* from the mouth of the Kamchatka River to 'Cape Chukotka,' from the Asian cape to the American cape and back as far as Karaginski Island. Also, there is the route of T. Perevalov's Expedition on this map (not earlier than 1767). The route goes from south to north between the first and second islands in Bering Strait and ends on the American Cape.[9] It was one of many attempts to show on maps the approximate route of Gvozdev's ship. Recent examples of some routes shown on historical maps are "Great Russian Geographic Discoveries in Pacific and North Oceans in the First Half of the 18th Century" from *Marine Atlas of 1958* and "Famous Voyages and Expeditions in Siberia from the end of 16th to the 20th Centuries" from *The History of Siberia* (Vol. 2).

Scientists of the eighteenth century appreciated the value of the many documents left by Gvozdev. For example, scientist M.V. Lomonosov said that they "demonstrate with certainty that the North Siberian Ocean connect the Atlantic Ocean with the Pacific Ocean and that Asia is separated from North America by water," and "that Gvozdev's documents also prove that there are islands and a land adjacent to the Chukotka Peninsula" (39, p. 451). After becoming Governor of Siberia, F.I. Somoinov communicated with Empress Elizabeth II his opinion of the existence of the New World. He, who used to meet with Gvozdev, wrote in his report of November 27, 1757: "The fact, that...in 1732 Gvozdev sailed from Okhotsk and he was near that land, though the journal of the expedition was not found, and he told me that he had seen the forests on that land, proving its existence" (25, p. 41).

Forty-six years after Gvozdev's expedition, the expedition of British mariner J. Cook saw the islands in the strait. Not giving every island a name, Cook called them all Diomede Islands (71, pp. 243-244); the northwest part of the American peninsula, "Cape Prince of Wales" (in the translation of the 18th and 19th century "Cape of Prince Valski") and the Island of Okiban (Ukivok), "King" Island. Academician P.S. Pallas, who compared the results of the voyages of 1732 and 1778, commented about the precise quality of the Russian maps. "The American coast adjacent to Cape Chukotka," he wrote, "is absolutely identical on Gvozdev's map and Captain Cook's description" (48, p. 142). In 1819, historian V.N. Berkh thought that Gvozdev had determined "the exact mathematical latitudes of the three small islands in Bering Strait and the American cape" (12, p. 10).

On the map of 18th century Russia, scientist and Professor A. Wilbreht does not mention the Diomede or Ukivok islands. On the map of 1784, prepared by Cossack I. Kobelev, is marked, "American cape seen by geodesist Gvozdev in 1732 and called Gigmalskaya Land." This name couldn't be found. "Gigmalskaya Land" comes from the Chukchi word "kihmiltsi," "Cape Kying-Myin," "Land Kying-Mying" etc. in the 18th century. All these words come from the Chukchi language. The Chukchi called America: Land Kyiyimyilyit (8, p. 128). On A. Wilbreht's map, published in 1787 in Russian and French, is written: "Cape Prince of Wales" and then the ironic notice "that the cape was known by Gvozdev in 1732 as Land Kyigmalskaya." The Soviet scientist blamed J. Cook for his appropriation of the name of the North American Cape (81, p. 9) and recalled that famous mariner V.M. Golovnin said how much foreign sailors liked to give names to new places. He wrote that if the Russians did the same during the 1830's and 1840's in America, then "all the capes and gulfs on the American coast would have names of Russian ministers, dukes and government officials" (18, p. 121). Though the name offered by Cook became widespread and remains today, some Russian scientists used to refer to it differently. For example, V.N. Berkh, on his 19th century maps, constantly called the American cape, "Gvozdev's Cape" or "the cape discovered by Gvozdev in 1732." The restoration of historical justice is very important: the cape has to be renamed and called after the person who first discovered it, Peter the Great's geodesist, M.S. Gvozdev.

The problem of reflecting and keeping Russian priority in names of many geographic locations troubled Russian scientists and seamen long ago. During examination of the Diomede Islands in 1791 G.A. Sarichev renamed them "Gvozdev's Islands."[10] The authors of *Russian Sailors* (1953) in their mentioning of Gvozdev's Islands (Diomedes) made an error and said that G.F. Müller called the islands so in the middle of the 18th century. Lt. O.E. Kotzebue continued research of Gvozdev's Islands in 1816 and called one of the islands after famous Russian mariner J.F. Kruzenshtern. Because of heavy fog he "discovered" a non-existing fourth island and named it after Captain-Lieutenant M.J. Ratmanov. When it became clear that the fourth island did not exist, G.S. Shishmarev transferred the name to the most westerly of Gvozdev's Islands in 1820. These geographic names remain after Kotzebue and Shishmarev as assigned by F. Bichi in 1826. He quite distinctly put on the map Ratmanov and Kruzenshtern Islands and he called the smallest island to the east, Fairway Rock (13, p. 110). Many errors have been made regarding these islands. For example, in 1828, F.P. Lütke, after examining Gvozdev's Islands, gave the old names back to every island and called them together, Diomede Islands, thinking V. Bering had called them so. But Lütke didn't know that the Captain-Commander and the crew of the *Gabriel* saw only one island, Ratmanov, in 1728. That is why the islands seen and described by Gvozdev in 1732 did not retain their names on Russian and foreign maps. Most often people call these islands Diomede Islands. But still one can find "Gvozdev's Islands" (French Atlas of A. Dufua of 1860, Swedish Atlas of Bonjer of 1951 and others) or some double names as on the map of L.A. Zagoskin in 1844 (31), "Gvozdev's or Diomede Islands," or in Marine Atlas 1950-1958, "Diomede Islands (Gvozdev's)." Famous Russian geographer P.P. Semyonov-Tyan-Shanski and academician L.S. Berg suggested naming the Diomede Islands after Gvozdev but it still has not happened (45, p. 37).

In spite of many attempts to undervalue the importance of the expedition of 1732 (76), modern scientific (73) and scientific popular literature (77) recognize the great achievements in the discovery of Northwest America by Russian seamen.

More than 250 years ago, on August 21, 1732 the crew of the ship *St. Gabriel* under the leadership of geodesist Mikhail Spiridonovich Gvozdev were the first Europeans to reach the Northwest American coast in the region of Seward Peninsula. Pioneers in exploring Bering Strait islands and the Alaskan coast made great geographic discoveries. That was the beginning of a new period in social-economic, political and trade relations between Russia and America, two nations divided by a narrow strait between two continents. They took the first steps on Russia's way to becoming a "Pacific Ocean Country."

The greatest geographical event in the first half of the 18th century was the expedition and discoveries by M.S. Gvozdev. As the result of the voyage of the *St. Gabriel* in 1732, the first correct map of Bering Strait, the islands in the strait (Diomede-Gvozdev Islands) and part of the Alaskan coast came into being. Comparing this expedition with that of Bering's First Kamchatka Expedition, it was much more successful in describing the northern part of the Pacific Ocean. Gvozdev's crew was the first to cross the strait from west to east and prove the separation of the Asian and American continents. In addition, they described on a map the correct location of Alaska and the shortest way there from the east (Cape Dezhnev) through the (Diomede-Gvozdev) islands. The members of the expedition were on one of the islands of the Bering Strait, collected considerable information about the Chukotka native people, the islands, the Big Land and also much physical geographical information about the region.

M.S. Gvozdev's reports-descriptions are the first known documents about the achievements of Russians on the Northwest American coast. Scientific results of the voyage, presented on the map of 1743, significantly influenced the development of map science and exploration of coasts and Bering Strait in the 18th and beginning of the 19th centuries. Many maps of the Second Kamchatka Expedition, Marine Academy in 1745-1746, World Map of G.M. Lovitsa in 1746, Javile's Map of Discovery of 1752, G.F. Müller's map of 1754-1758, F.I. Somoinov's map of 1760, T. Perevalov's map of 1767, I.F. Truskott's - G.F. Müller's map of 1773, I. Kobelev's map of 1779, A. Wilbreht's map of 1787, G.A. Sarichev's map of 1791, O.E. Kotzebue's map of 1816, and V.N. Berkh's map of 1821 reflect geographical discoveries made by Gvozdev. Although Gvozdev's discoveries became known and recognized during his lifetime, Berkh, the historian and traveler, said about Gvozdev, "His name will not be forgotten as long as the islands described by him exist" (12, p.10). In Gvozdev's honor, a cape on Sakhalin Island bears his name.

The information of Gvozdev's voyage was researched and used by all the expeditions to the northeast that followed, particularly the Second Kamchatka Expedition. Many participants of the voyage of 1732 became members of the crews of the *St. Peter* and *St. Paul* that, under the leadership of Bering and Chirikov, discovered additional portions of the northwest American coast and Aleutian Islands.

Geodesist Gvozdev also served in the Second Kamchatka Expedition. Accordingly, a symbolic note could be placed on many maps of different geographical places stating, "Geodesist Gvozdev was here." Among them would be the sea coast near Okhotsk described by Gvozdev in April-June 1741, the mouth of the Bolshaia River, Kurile Islands, the east coast of Sakhalin Island, Uda River and Shantar Islands examined by Gvozdev with Shelting and Rtishev during their voyages on the double-sloop *Nadezhda* in September-October 1741 and July-October 1742; and various regions of Irkutsk which Gvozdev described until the last days of his service in Siberia.

Footnotes to Chapter 7

1. TsGADA, f. 248, Map No. 1911.
2. TsGADA, f. 192, No. 36; TsGAVMF (46, No. 69, 70).
3. TsGADA, f. 192, No. 50.
4. TsGAVMF, f. 230, Case 1, p. 56.
5. Ibid., other side of p. 61; f. 216, Case 48, pp. 742 and other side.
6. Ibid., f. 172, Case 408, pp. 1, 46-49a, 413-other side of 416; f. 913, Description 1, Case 1, other side of pp. 84 - 92.
7. Ibid., f. 216, Case 73, pp. 111-113, 126-128.
8. Ibid., pp. 115 and the other side of 128.
9. TsGADA, f. 192, No. 10.
10. In the note about the Gvozdev (Diomede) Islands, the authors of "Russian Sailors" (1953) erroneously state that these islands were so called by G.F. Müller in the middle of 19th Century.

Dates in the Life of M.S. Gvozdev

1715 December 24. Mikhail Gvozdev was sent to Moscow Mathematical-Navigational School.

1716 January - 1718. Student in Moscow Mathematical-Navigational School.

1719-1721 Student in Marine Academy in St.Petersburg.

1721-1725 September. Worked on the project of description and preparation of maps of Novgorod District, as a member of a detachment under the leadership of Major General M. Ya. Volkov.

1725 September - 1727 April. Continued his studies at the Marine Academy.

1727 May 20. Geodesist in the Far East Expedition of A.F. Shestakov.

1727 August - 1728 June 29. Trip on the route of St. Petersburg-Moscow-Tobolsk-Yakutsk.

1729 August - beginning of October. Trip from Yakutsk to Okhotsk.

1730 September 30. Arrived at Kamchatka and spent the winter in Bolsheretsk Ostrog.

1731 June 23 - July 9. Voyage on the *St. Gabriel* from Bolsheretsk to the mouth of the Kamchatka River.

1732 February 11. Promoted to commander of the Kamchatka Detachment of the D.I. Pavlutski Expedition and leader of the voyage to the Big Land.

1732 May 1 - 1733 November 30. Commander of Kamchatka Detachment.

1732 July 23 - September 27(28). Voyage on the ship *St. Gabriel* from Lower Kamchatka Ostrog to North American coast and return.

1732 August 21. Russian sailors under Gvozdev's leadership were the first Europeans to reach the Alaskan coast in the region of Cape Prince of Wales.

1733-1735. Managing the construction of the Lower Kamchatka Ostrog.

1735-1738. Trial in Tobolsk regional administration and his imprisonment.

1738 June - 1739 May. Trip from Tobolsk to Irkutsk to the location of his new assignment.

1739-1741 July 14. Geodesist in Port of Okhotsk.

1740 December. Research and preparation of the map and description of the Okhotsk - Malchikan River region.

1741 April - July. Official trip to Uda river and research in the Sea of Okhotsk.

1741 July 14. Geodesist of the Second Kamchatka Expedition Spanberg's detachment).

September 4 - October 14. Continued research of the Sea of Okhotsk on the double-sloop *Nadezhda* and description of Uda River.

1742 May 2. Assistant to Midshipman of the double-sloop *Nadezhda* from Bolsheretsk to Kurile Islands, Amur River and Okhotsk.

1743 July - August. Geographic and hydrographic research in the regions of the Okhota and Malchikan rivers.

1743 September 9 - October 8. Completed the first map (with K. Yushin and E. Rodichev) of his voyage in 1732 on the ship *St. Gabriel* to North American coast.

1744 June 13 - August 13. Voyage with A. Shelting's crew on the rivers from Okhotsk to Yakutsk to Eniseisk.

1744 September - 1754 December. Geodesist of Admiral's detachment of Tomsk's garrison.

1749 Promoted to Geodesist Lieutenant (retroactive to 1732).

1754 December 9. Map researcher in Irkutsk Regional Administration.

1755 April 5 - 1758. Worked in Irkutsk District making descriptions and preparing maps of agricultural regions.

1759 Senate's order concerning retirement of M.S. Gvozdev.

Documents by M.S. Gvozdev[1]

1721-1725. Drawings and descriptions of rivers and "various places" of Novgorod District.

1727 June 20. List of necessary geodesic instruments and materials for map preparation in expedition of A.F. Shestakov (TsGADA, f. 248, Book 690, pp. 307-308).

1727 July 15. Explanation of the purposes of various instruments of geodesic work (TsGADA, f. 248, Book 690, pp. 309-311).

1731 October 18. "Written Report" about activity of 1731 (with I. Speshnev) (TsGADA, f. 248, Book 666, pp. 298-303).

1732 May 1. Notice to navigator Ya. Gens about the appointment of the geodesist as leader of the voyage to the Big Land (TsGAVMF, f. 216, Case 5, pp. 126-127).

1732 May 3. Description of the ship *St. Gabriel*, its equipment, accounting reports and roster of personnel in Gvozdev's detachment (TsGAVMF, f. 216, Case 3, pp.188-195).

1732 May 2 - July 13. Various administrative orders during the preparation of the *St. Gabriel* for the voyage to the Big Land (TsGAVMF, f. 1002, case 1, pp. 18-20).

1732 July 2. The letter to navigator Ya. Gens concerning the departure of the voyage to the Big Land (TsGAVMF, f. 246, Case 4, p. 41).

1732 July 23 - September 27(28).* Log of the ship *St. Gabriel* (completed with assistant navigator I. Fedorov).

1732 November 10.* Letter to assistant navigator I. Fedorov about the necessity of creating a log and map of voyage of 1732.

1732 December 19.* Report to Anadyr's commander D.I. Pavlutski concerning the voyage in 1732 to the North American coast and a copy of the ship's log.

1733 June 22.* Report of Commander of Okhotsk Port, G.G. Skorniakov-Pisarev concerning the voyage of 1732 to the coast of North America and the original ship's log, received December 23, 1733.

1733 Various personal documents from Kamchatka (TsGAVMF, f. 216, Case 4, pp. 180 and opposite, Case 5, pp. 7 and opposite).

1738 June 13. Report to Siberian Regional Administration concerning service in the Shestakov-Pavlutski Expedition (TsGAVMF, f.216, Case 24. pp. 599-600, 617-618) Published in 1956 (25, pp. 45-6) under the incorrect title, "Report of M.S. Gvozdev given to Siberian Regional administration concerning the voyage to the American coast.

1740 July.* The request of Empress Anna Ivanovna and Admiralty College concerning a promotion with a description of service completed since 1721.

1740 November - December.* "Descriptions of possible living places" and the drawing of the region of the Malchikan and Okhota rivers. In archive catalogs of 1852 and 1917 of the Hydrographyal Department there is "The plan of Okhota river from the point which is 60 versts from the river's mouth, signed by Gvozdev (about 1740) in Russian language" (Catalog of Atlases and Maps of the Main Hydrographical Department of Marine Ministry. Ch. II, p. 1917, No. 966. Catalog of Atlases and Maps of Archives Hydrographical Department of Marine Ministry, Ch. 2, SPB, 1852, p. 238).

1740-1741.* "Two plans convenient for living quarters in Okhotsk, prepared by Gvozdev." According to the Admiralty College, 1754 (TsGAVMF, f. 216, Case 73, pp. 115 and opposite 128).

1741 April 13. Report to A. Devier, commander of Okhotsk Port, concerning the voyage of 1732 to the coast of North America. Published in 1850 (51, pp. 397-401) and 1948 (30, pp. 236-243).

1741 April-July.* Description of the coast from Okhotsk to Uda River. Map and journal of the voyage.

1741 September 4 - October 14. Journal from the double-sloop *Nadezhda* on voyage from Okhotsk - Shantar Islands – Kurile Islands – Bolsherestsk Ostrog (with V.A. Rtishev). (TsGAVMF, f. 913, Des.1, Case 38, pp. 440 to other side of 460)

1741 September - October.* Description of the Uda River and Shantar Island with A.E.Shelting.

1741 November. * Map of voyages made by *Nadezhda* (with A.E. Shelting).

1742 July 22 - October 13. Journal of *Nadezhda* on voyage to Kurile Islands, to east coast of Sakhalin, to Okhotsk (with V.A. Rtishev) (TsGAVMF, f. 913, Description 1, Case 38, pp. the other side of 495-521).

1742 November. "Map of the route Penzhin Sea - Bolsheretsk – Kurile Islands, prepared by midshipman Rtishev and geodesist Gvozdev"[2] according to the Admiralty College in 1754 (TsGAVMF, f. 216, Case 73, pp. 111-113, 126-128).

1743 January 12. Report of officers commission concerning soldier M. Polujanovsky vs. A.E. Shelting and P. Belousov (with officers of the Second Kamchatka Expedition V.A. Rtishev, A. Korostelev, A. Shaganov, and I. Bobovksy) (TsGAVMF, f. 216, Case 53, pp. 521 and the other side).

1743 July 7. Report to A.E. Shelting concerning the economic uselessness of building a dockyard on the Malchikan River (with V.A.Rtishev and Yu. Aritlander) (TsGAVMF, f. 126, Case 53, pp. 60-61) Published in 1979 (52, pp. 234-5.).

1743 August 24. Autobiography sent to Empress Elizabeth Petrovna. (TsGAVMF, f. 216, Case 56, pp. 966-967) Published in 1956 (25, pp. 43-45).

1743 September 1. Report to Captain M.P. Spanberg concerning the voyage to the coast of North America. (TsGAVMF, f. 216, Case 53, pp. 733 to the other side of 738).

1743 October 8.* Portions of the private diary of I. Fedorov concerning the voyage of the *St.Gabriel* in August 1732 (with K. Yushin and E. Rodichev).

1743 October 8. "The map from the Kamchatka River to Chukotka and to the islands" (with K. Yushin and E. Rodichev) sent to Irkutsk Regional Administration on November 5, 1743 (TsGAVMF, f. 216, Case 53,the other side of p. 742).

1743 November 4.* Two letters (to the Senate and Admiralty College) concerning promotion with a description of service activity and achievements.

1747 February 12.* Letter to the Admiralty College concerning granting the geodesist the next rank.

1747 July 5. Same as above.

1758 July. Letter to Empress Elizabeth Petrovna concerning retirement. (TsGADA, f. 248, Description 113, Case 485a, pp. 399 and the reverse of 400) Published in 1979 (52, pp. 155-159).

Footnotes

1 * These documents are missing but are known in historical sources by reference and interpretation.
2 Mentioned in archive catalogs of 1852 and 1917 of Hydrographical Department under the title "Map of the voyage from Bolsheretsk, along Kurile Islands prepared under the supervision of Spanberg by midshipman Rtishev and geodesist Gvozdev in 1742 in Russian." See Evald B. Bibl, Part II, No.1033, Catalog of Atlases, 1852, p. 247. Published in 1851, (59, pp. 88-103), in 1948 (30, pp. 244-248) and in 1979 (52, pp. 150-154) translated into English in 1914 (75, pp. 160-162) and in 1922 (74, Vol. 1, pp. 22-24)

Bibliography

1. Alexander, B.V., "Opisanie rykopisnykh kart XVIII v., khraniashchihksia v otdele rukopisoi knigi Biblioteki Akademii nauk USSR" Description of Hand Drawn Maps of the Eighteenth Century" In: Gnucheva V. F. *Geograficheskii departament Akademii nauk XVIII*. [*Geographical Depatment of the Academy of Sciences in the Eighteenth Century*]. Moscow and Leningrad, 1946.
2. Alekseev, A.I., *Syny otvazhnye Rossii*. [*Courageous Sons of Russia*]. Magadan, 1970.
3. Andreev, A.I., *Ocherki po istochnikovedeniiu Sibiri XVIII v.* [Essays on source materials for Siberia, the 18th century]. Moscow - Leningrad, 1965.
4. Andreev, A.I., *Zametki po istoricheskoi geografii Sibiri XVI – XVIII vv.* [*Notes on the Historical Geography of Siberia in the XVII – XVIII Centuries.*] Izv. VGO, 1940, 2nd Edition.
5. *Atlas geograficheskikh otkrytii v Sibiri v severozapadnoi Amerike, XVII-XVIII vv.* [Atlas of geographical discoveries in Siberia and in northwestern America, 17th-18th centuries]. Moscow: Nauka, 1964.
6. Belov, M.I., *Arkticheskoe moreplavanie s drevneishikh vremen do sereediny XIX veka.* [Arctic seafaring from ancient times to the middle of the of the ninteenth century], Ia.Ia. Gakkel, A.P. Okladnikov, and M.B. Chernenko, eds, Moscow: Morskoi transport [Istoriia otkrytiia i osvoeniia severnogo morskogo puti, 1], 1965.
7. Belov, M.I., "O sostavlenii Generalinoi karti Vtoroi Kamchatskoi ekspeditsii 1746 g." [About the preparation of the map of the Second Kamchatka Expedition in 1746]. In *Geograf*, Moscow – Leningrad, 1954, Vol. 3.
8. Berg, L.S., "Ievectiia o Beringovom prolive i ego beregakh do Beringa I Kuka" [Information about Bering Strait and its coasts prior to Bering and Cook], In *Zapiski po gidrografii*, Vol. 2., No. 2, 1920.
9. Berg, L.S., *Otkrytie Kamchatki i ekspeditsii Beringa, 1725-1741.* [The discovery of Kamchatka and the Bering Expediton 1725-1741] 3rd Edition, Moscow – Leningrad, 1942.

10. Berkh, V.N., "Raznye izvestiia i pokazaniia o Chukotskoi zemle" [News of Chukotka and the North], *Cev. Arkh.*, 1825, Part 18, No. 22.

11. Berkh, V.N., *Pervoe morskoe puteshestvie rossiian, predpriniatoe dlia resheniia geograficheskoi zadachi: Soediniaetsia li Axiia s Amemikoi? i sovershennoe v 1727, 28 i 29 godakh pod nachalstvom Flota kapitana 1-go ranga Vitusa Beringa.* [The first Russian maritime voyage, undertaken to decide the geographical question: Is Asia joined to America? and completed in 1727, 1728, and 1729 under the command of fleet captain of the first rank Vitus Bering]. St. Petersburg: Imperatorskaia akade. nauk, 1823.

12. Berkh, V.N., "Puteshestvie golovy Afanaciia Shestakova I pokhod maiora Pavlutskogo v 1729 i 1730 godakh," [The journey of Cossack Head Afanasiy Shestakov and the voyage of Major Pavlutski in 1729-1739]. *Syn otechestva*, No. 20, 1819.

13. Birkengof, A.L., "K istorii nazvanii 'ostrova Gvozdeva (Diomid),' 'Beringov piroliv,' i 'Beringovo more,'" [On the history of the names "Gvozdev Island (Diomede)," "Bering Strait," and "Bering Sea"]. In *Strany Vostoka*, No. 13. 1972.

14. Vdovin, I.S., *Ocherki istorii i etnografii Chukchei* [Essays on the history and ethnography of the Chukchi]. Moscow and Leningrad: Nauka, 1965.

15. Vdovin, I.S., "Chertezhi Chukotki 1742 i 1746 gg." [Drawings of Chukotka: 1741 and 1746], *Izv., VGO*, 1943, 4th Edition.

16. Vize, V.Yu., *Moria Sovetskoi Arktiki.* [*Seas of the Soviet Arctic.*] Moscow - Leningrad, 1948.

17. Voskoboinikov, V.N., *Slovo na karte: Iz istorii geograficheskikh nazvanii Kamchatskogo poluostrova* [Words on the Map: the History of the Geographical Names of Kamchatka]. Petropavlovsk – Kamchatka, 1962.

18. Golovnin, V.M., *Sochineniia i perevody, tom III* [Works and translations, Vol. III]. St. Petersburg, 1864.

19. Goldenberg, L.A., "Karta Shestakova 1724 g." [Shestakov's map of 1724], In *Isbolzovanie starykh kart v geograficheskikh i istoricheskikh issledovaniiakh.* Moscow, 1980.

20. Goldenberg, L.A., "Tobolskaia opis V. Shishkina 1738 g. i novye istochniki o pokhodakh na Chukotku, Kamchatku i k beregam Ameriki" [The Tobolsk inventory of V. Shishkin, 1738, and new sources on voyages to Chukotka, Kamchatka and the coasts of America], *Sovetskie Arkhivy*, No. 3, 1980.
21. Goldenberg, L.A., *Katorzhanin sibirskiy gubernator. Zhizn i trudy F.I. Soymonova* [Convict and Siberian Governor. The life and works of F.I. Soimonov]. Magadan, 1979.
22. Goldenberg, L.A., "Geograficheskii departament Akademii nauk i sazdanie pervogo Akademicheskogo atlasa *(1738-1799 gg.)* [The Geographical Department of the Academy of Sciences and the creation of the first academic atlas (1738-1799)]. In Ocherki istorii geograficheskoi nauki v SSSR. Moscow: Nauka, 1976.
23. Grekov, V.I., *Ocherki iz istorii russkikh geograficheskikh isledovanii v 1725-65 gg.* [Essays on of the history of Russian geographical exploration, 1725-65]. Moscow: Nauka, 1960.
24. Divin, V.A., *Russkie moreplavaniia na Tikhom okeane v XVIII veke* [Russian Seafaring on the Pacific Ocean in the eighteenth century]. Moscow: Mysl., 1971.
25. Divin, V.A., *K beregam Ameriki; plavaniia i issledovaniia M. S. Gvozdeva, pervootkryvatelia Severo-Zapadnoi Ameriki* [To the shores of America; the navigations and explorations of M.S. Gvozdev, the first discoverer of northwestern America]. Moscow: Gos. Izd-vo greor. Lit-ry, 1956.
26. Divin, V.A., *Velikiy russkiy moreplavatel A.I. Chirikov* [The great Russian seafarer A.I. Chirikov], Moscow: Geografizgat, 1953.
27. Efimov, A.V., *Iz istorii velikikh russkikh geograficheskikh otkrytii* [History of great Russian geographical discoveries], Moscow: Nauka, 1971.
28. Efimov, A.V., "Iz istorii kartografii Dalnego Vostoka" [The history of cartography in the Far East], In *Sbornik statei po istorii Dalnego Vostoka*. Moscow, 1958.
29. Efimov, A.V., *Iz istorii velikikh zusskikh geograficheskikh otkrytii v Severnom Ledovitom i Tikhom okeanakh, XVII-pervaia polovina XVIII v* [From the history of great Russian geographical discoveries in the northern Arctic and Pacific oceans, 17th century to the first half of the 18th century]. Moscow: Geografizdat, 1950.

30. Efimov, A.V., *Iz istorii russkikh ekspeditsii na Tikhom Okeane Pervaia polovina XVIII veka* [From the history of Russian expeditions to the Pacific Ocean, first half of 18th century Moscow, 1948.
31. Zagoskin, L.A., *Puteshestviia i issledovaniia leitenanta Lavrentiia Zagoskina v russkoi Amerike v. 1842-1844 gg.* [The voyages and research of Lt. Lavrenty Zagoskin in Russian America in 1842-44]. Moscow, 1956. *Gidrograficheskogo departmenta* St. Petersburg, 1851, 9:78-107.
32. Znamenski, S.V., *V poiskakh yaponii* [Searching for Japan]. Vladivostok, 1929.
33. Kirilov, J.K., *Tsvetushchee sostoianie Vsepossiiskogo gosudarstva* [The wealthy condition of the Russian Empire]. Moscow, 1977.
34. "Kolonialnaia politika tsarizma na Kamchatke i Chukotke v XVIII v." [Colonial policy of Czarism on Kamchatka and Chukotka in the 18th century]. In *Sbornik arkhivnykh materialov*, Leningrad, 1935.
35. Kotzebue, O.E., *Puteshestvie v yuzhnyi okean i v Beringov proliv dlia otiskaniia severovostochnogo morskogo prokhoda, predpriniatoe v 1815, 1816,1817 i 1818 godakh na korablie Riurikie* [Voyage to the Pacific Ocean and Bering Strait in Search of a Northeast Passage in 1815, 1816, 1817 and 1818 of the ship *Rurik*], St. Petersburg 1821, Part I.
36. Krasheninninkov, S.P., *Opisanie zemli Kamchatki* [Description of the land of Kamchatka]. Moscow - Leningrad, 1949.
37. Lapter, S.N., "O nekotorykh petrovskikh geodezistakhtopografakh, uchastnikakh Vtoroi Kamchatskoi ekspeditsii" [About some Geodesists participating in the Second Kamchatka Expedition], In Uchen. zap. perm. un-ta, 1967, No. 615.
38. Lütke, F.P., *Puteshestvie vokrug sveta na voennom shliupe Seniavin v 1826-29 godakh.* [Voyage around the world on the sloop-of-war *Seniavin*, 1826-29]. St. Petersburg, 1835.
39. Lomonsov, M.V., "Kratkoe opisanie raznykh puteshestvii po severnym moriam I pokazanie vozmozhnogo prokhodu Sibirskim okeanom v Vostochnyi" [A Short description of various voyages to the Northern Sea and possible route via the Siberian Ocean to the

Pacific Ocean]. In *Polnoe sobranie sochinenie*, Moscow – Leningrad, 1952, Vol. 6.

40. "Lotsiya Severozapadnoi Chasti Vostochnogo okeana" [Map of and sailing directions of Northwest portion of the Pacific Ocean]. St. Petersburg, 1909, Part 4, *Beringovo more i proliv Beringa*.

41. *Materialy dlia istorii russkogo flota* [Materials for a history of the Russian Navy]. St. Petersburg, Part 3, 1866, Part 5, 1875.

42. Medushevskaya, O.M., "Kartograficheskie istochniki po istorii russkikh geograficheskihk otkritii na Tikhom okeane vo 2-i polovine XVIII veka." [Cartographic sources on the history of Russian geographical discoveries in the Pacific Ocean in the second half of the 18th century]. *Trudy Moskovskogo gosudarstvennogo istoriko-arkhivnogo instituta*, Vol. 7, 1954.

43. Müller, G.F., "Opisanie morskikh puteshestvii po ledovitomu i po vosto chnomy moriu s rossiiskoi storony uchenennykh" [Description of the ocean voyages made in the Arctic and Pacific oceans from the Russian side], *Sochineniia I perevody*. St. Petersburg: Imperatorskaia Akademiia nauk, 1758, Jan-May; Vol. VII, July-Sept., Nov; Vol. VIII.

44. *Sea Atlas*, 1950-1958

45. Naumov, A.V., "O kapte russhikh otkeyfii XVIII veka a severnoi chasti Ameriki" [About the map of 18th century Russian discoveries in the northern part of America], *Trudy Moskovskogo instituta inzhenerov geodezii, aerosiomki i kartografii*, No. 13, 1951

46. *Opisaniye starinnyukh atlasov, kart i planov XVI,XVII, XVIII vikov i poloviny XIX veka, khranyashchikhsya v Arkhive Tsentralnogo kartograficheskogo proizvodstva VMF* [Description of old atlases, maps and plans from the 16th, 17th, 18th and early 19th centuries, reserved at the Archives of the Central Cartographic Establishment of the Navy]. Leningrad, 1958.

47. Orlova, E.P., "Chertizhi Chukotki Iakova dimdermana i Timofeia Perevalova" [Drawings of Chukotka by Jacob Linderman and Timothy Perevalov], in *Geografii Dalnego Vostoka*, 1957, 3rd Ed.

48. Pallas, P., "O possiiskikh otkrytiiakh v moriakh mezhdy Aziri i Amerikoi" [Russian discoveries in the seas between Asia and America] In: *Mesyatseslov istoricheskiy i geograficheskiy na 1781 god*. St. Petersburg, 1781.

49. Polevoy, B.P., *Pervootkryvateli Sakhalina* [First Discoverers of Sakhalin]. Iazhno-Sakhalinsk, 1959.
50. "Polnoe sobvranie zakonov Rossiiskoi imperii" [Complete collection of the laws of the Russian empire], St. Petersburg, 1830-1884.
51. Polonski, A.S., "Pokhod geodezista Mikhaila Gvozdeva v Beringov proliv, 1732 goda" [The journey of geodesist Mikhail Gvozdev in the Bering Strait, 1732], *Morskoi sbornik* St. Petersburg, 4, no. 11:389-402.
52. *Russkaya tikhookeanskaya epopeya* [The Russian Sea Epic]. Khabarovsk, 1979.
53. *Russkie otkrytiia b Tikhom okeane i Severnoi Amerike v XVIII veke*. [Russian discoveries in the Pacific Ocean and America in the 18th century]. Moscow, 1948.
54. Sarichev, G.A., *Puteshestvie flofa kapitana Saricheva po Severovostochnoi Chasti Sibiri, Ledovitomu moriy i Vostochnomu okeanu, v prodolzhenii osmi let, pri geograficheskoi i astronomicheskoi morskoi ekspeditsii, byvshei pod nachalstvom flota kapitana Billingsa s 1785 po 1793 god.* [The voyages of Captain Sarichev's fleet in Northeast Asia, Arctic Ocean and East Ocean during the eight years with the Geographical and Astronomical Expedition under the leadership of Captain Billings, 1785 to 1793]. St. Petersburg, 1802, Parts 1 and 2.
55. Sgibnev, A.S., "Materialy dlia istorii Kamchatki: ekspeditsiia Shestakova" [Material for the history of Kamchatka: Shestakov's expedition], *Morskoi sbornik* St. Petersburg, 1869, Vol. 100, No. 2.
56. Sgibnev, A.S., "Istoricheskii ocherk glavneishikh sobytii v Kamchatke 1650-1856" [Historical outline of the principle events in Kamchatka, 1650-1856], *Morskoi sbornik* St. Petersburg, 1869, Vol. 101, Vol. 102, No. 6.
57. Sgibnev, A.S., "Navigatskie shkoli v Sibiri" [The Navigational Schools of Siberia]. *Mor.* St. Petersburg, 1866, Vol. 87, No. 11.
58. Sergeev, V.K., "Moskovskaia matematiko-navigatskaia Shkola" [Moscow Mathematical-Navigational School]. *Vopr. Geografii*, 1954.

59. Solokov, A.P., "Pervyi pokhod russkikh k Amerike, 1732" [The first Russian journey to America, 1732], *Gidrograficheskogo departmenta* St. Petersburg, 1851, 9:78-107.
60. Sokolov, A.P., "Severnaia ekspeditsiia, 1733-43 goda" [The Northern Expedition, 1733-43], *Gidrograficheskogo departmenta* St. Petersburg, 1851, Part 9.
61. Strelov, E.D., *Akty arkhivov Yakutskoi oblasti (1650-1800)* [Edicts form the Archives of the Yakutsk District (1650-1800)]. Yakutsk, 1916, Vol. 1.
62. Troitskii, S.M., "Sibirskaia administratsiia v seredine XVIII v." [The Siberian Administration in the middle of the eighteenth century]. In *Voiprosy istorii Sibiri dosovetskogo perioda*. Novosibirsk, 1973.
63. Fedorova, S.G., *Russkoe naselenie Aliaski i Kalifornii* [The Russian population of Alaska and California]. Moscow: Akademiia Nauk SSSR, 1971.
64. Fedorova, S.G., "K voprosu o rannikh russkikh poseleniiakh na Alaiske." [On the question of early Russian settlements in Alaska] In *Letopis Severa*, Vol. 4, 1964.
65. Fel, S.E., *Kartografiia Rossii XVIII veka* [Russian cartography of the eighteenth century]. Moscow: Geodezicheskaia literatura, 1960.
66. "Chukchi I zemlia ikh c otkrytiia etogo kraia do nastoiashchego vremeni" [The Chukchi and their region since discovery of that region until this date]. Magadan, vnutr. del, 1851, Sec. 34.
67. Shibanov, F.A., "Geodezicheskoe obrazovanie v Rossii v pervoi polovine XVIII veka." [Geodesical education in Russia in the first half of the eighteenth century]. In *LGU*, 1980, No. 6. Ser. Geologiia, geografiia, vyp 1.
68. Shibanov, F.A., *Ocherki po istorii otechestvennoi kartografii* [Essays on Domestic Cartography]. Leningrad, 1971.
69. "Ekspeditsiia Beringa" [Bering's Expedition]. Documents prepared for publication by A. Pokrovskii, Moscow, 1941.
70. *Atlas Homannianus*. Nuremberg, 1773.
71. Cook, J., *Voyage to the Pacific Ocean in the years 1776, 1777, 1778, 1779 and 1780*. Vol. II, London, 1785.

72. Coxe, W., *Account of the Russian discoveries between Asia and America*. London, 1780.
73. Fisher, R., *Bering's voyages. Whither and Why*. Seattle – London, 1977.
74. Golder, F., *Bering's Voyages*. New York, 1922, Vol. I, II.
75. Golder, F. *Russian Expansion on the Pacific, 1641-1850*. Cleveland 1914.
76. Hulley, C., *Alaska, 1741-1953*. Portland, 1953.
77. Lower, A., *Ocean of destiny: a concise history of the North Pacific, 1500-1978*. Vancouver, 1978.
78. Müller, G.F., *Voyages from Asia to America*. London, 1761.
79. Pallas, P., "Geographische Beschreibung des Anadyrflusses" In *Neue Nordische Beitrage*. St. Petersburg–Leipzig, 1781, Bd. 1, St. 2.
80. Sailing directions (Onroute). For the East coast of the USSR Hydrographic Center, Pub., 1972, N 155.
81. Svet Ya. and Fedorova S., "Captain Cook and the Russians" In *Pacific Studies*, 1978, Vol. II, No. 1.
82. United States coast pilot. No. 9. Pacific and Arctic coasts of Alaska, Cape Spencer to Beaufort Sea. Seventh Edition, 1964.

Abbreviations

AAN	The Archives of the Academy of Sciencc
AVPR	The Archives of the Foreign Policy of Russia
VMF	The Archives of the Central Cartographic Production of the Navy
VGO	All Union Geographic Society
F3	The Geographic Department of the Academy of Science
F21	The Cases of the Sea Ministry
F24	Sibirskii Prikaz and the Management of Siberia
F172	The Office of I.B. Chernyshov
F176	The Admiralty Office
F192	The Cartography Department of the Library of Moscow Main Archives of the Ministry of Foreign Affairs
F199	G.F. Müller ("portfolios")
F212	The Admiralty Board
F214	Sibirskii Prikaz
F216	Vitus Bering
F230	The Expedition Office of Admiral N.F. Golovin
F233	F.M. Apraskin
F248	The Senate
F339	The Russian-American Company
F913	The Archives of Hydrography
F1002	The Kamchatka Land Surveying Detachment
TsGAVMF	The Central State Archives of the Navy of USSR
TsGADA	The Central State Archives of Ancient Deeds

Glossary

Alexandria paper: type of paper used for maps

Altyn: a silver coin worth 6 denga or 3 kopecks

Arshin: .71 meters

Bidaraka: skinboat capable of carrying 20 people

Cossack: men who served the government for certain privileges such as not paying taxes

Denga: half of a kopeck

Destei: 24 pages of paper

Kukhta: one man skinboat

Piatidesiatnik: leader of 50 men

Pud: 36 pounds

Ruble: 100 kopecks

Sazhen: 1.83 meters

Stopa: 20 destei or 480 pages of paper

Verst: 3,500 feet or 1.06 kilometers

Yurt: semi-subterranean dwelling

Zolotnik: 4.26 grams

History of the Names of the Islands in Bering Strait

Source	Ratmonov	Diomede Kruzenshtren
1. Map "Kamchadalskaya land Lamsk and Penzhin seas, as were found and inspected as a result of various expeditions by Russian Cossacks and hunters for sable on water and land." (translated from German) From Homman's Atlas, 1725.	"Empty"	"Island where people live"
2. "Map representing Anadyr ostrog and Anadyr Sea" I. Lvov, not earlier that 1727-8 TsGADA, f, 192, Irkutsk Province, #26.	"On this island live people, who the Chukchi call ahuhalyat"	"On this island live people who Chukchi call peekeli"
3. "Index to the cities and famous Siberian places..." First Kamchatka Expedition, 1728. *Bering's Expedition*, p. 66.	"Island of St. Diomede" 67°00'N.L. 125°42'E. of Tobolsk	
4. Reports of Geodesist M.S. Gvozdev 1741 and September 1, 1743, Efimov, 1948, pp. 236-243 and pp. 244-249.	"First Island"	"Second Island"

Islands		
Fairway Rock	King Island	Cape Prince of Wales
-	-	Not named (on a different version of the map: "Incognita.")
-	-	" Land is Big, and there live people, who Chukchi call kiginedyat."
-	-	
Not named	"Fourth Island"	"Big Land"

Source	Ratmonov	Diomede Kruzenshtren
5. Map of North-East Asia. Y.I. Lindenau, 1742. TsGADA, f. 248, Maps, No. 1910.	Without a name	Without a name
6. "Map Mekatorskaya from Okhotsk to the Chukota Nos...." M.P. Spanberg's Oct. 1743 TsGAVA, f. 192, VUA No. 23431; ATsKP VMF No. 69, 70.	"St. Diomede"	Without a name
7. "Map of Northern part of the East Ocean with Bering Strait and part of the Arctic Ocean" Sea Academy, 1746. ATsKP, VMF, No.82; see TsGVEA, f. VUA, No. 23466, 20227.	"St. Diomede"	Without a name
8. "Mappa Monde..." G.M. Lovitsa, 1746 from "Atlas von Hundert Carte Atlas Humonnianus, 1773.	-	
9. Map of Chukotka peninsula and part of Perevalov's Kamchatka, October 1754 Atlas, 1964, No. 124	"People here"	Island of large-toothed Chukoch "On this island live people called large-toothed peekeli"

Islands		
Fairway Rock	King Island	Cape Prince of Wales
Without a name	Without a name	"Big Land"
Without a name	Without a name	"Geodesist Gvozdev was here, 1732"
Without a name	"Nomand's tent"	(in the legend) "Land, laid across from Chukotka East corner to the east laying from 65° to 68° N.L. taken from the map of geodesist Gvozdev."
-	-	"detecta a Gwosdew 1730"
-	-	Island or Big Land people live here who Chukchi call kiginedyat."

Source	Ratmonov	Diomede Kruzenshtren
10. Map of G.F. Müller, 1754-1758 "Nouvelle carta des decouvertes, faites par des vaisseaux Russiens aux cotes inconnues de l'Amerique Septentrionale avec les pays adjacents" TsGADA, f. 192, Irkutsk Province, No. 50.	" I. St. Diomede	-
"A Map of the Discoveries made by the Russians on the Northwest coast of America" London, 1761.	The same	-
11. Map of Northeastern Asia pointing out existing and planned sea and land routes. F.I. Soimonov, 1760 TsGADA, f. 248, maps, No. 1911 Rough draft of this map in TsGVEA, Efimov, 1958, il. 4.	"Diomede Island"	-
12. "Map of Province of Yakutsk, Chukota land, Kamchatka land with nearby places and part of America with nearby islands" Perevalov's, not earlier than 1763 "Red Archives," 1936 T. 1, p. 160.	"Magli Island. Large-toothed Chukchi live here"	"Igali Island Large-Toothed Chukchi live here"

Islands		
Fairway Rock	King Island	Cape Prince of Wales
-	-	"Cote decouverte par le geodesiste Gwosdew 1730"
-	-	"Coast Discovered by geodesist Gvozdev in 1730"
-	-	"Land western part of America, to which in 1732 Gvozdev has been to, the exactness of which is not known"
Without a name	Without a name	"Part of North America. Live here are people called kuhkmults

Source	Ratmonov	Diomede Kruzenshtren
13. "Map of Chukota Land with part of American shore" Daurkin, 1765. TsGVEA, f. VUA, No. 23435.	(Abridged) "Imyuaglin Island. There large toothed people live"	(Abridged) "Inyallin Island. There live same people as on Imyuaglin"
14. "Map of Eastern Asia. Vertlugov, 1767. Coxe, 1780, p. 322; Alexandrov, No. 104.	Without a name	Without a name
15. "Map of Yakutsk Province, Kamchatka and Anadyr River with nearby Chukota land, also part of N. America and Japan." Perevalov, not earlier than 1767. TsGADA, f. 192 Irkutsk Prov #10.	"Magli Island. Large toothed Chukchi live here"	"Ogalgi Island. Large toothed Chukchi live here"
16. "Map of Northeastern Asia and North America" F. Plenisnera, 1770 Alexandrov #123, Atlas #131.	"First Inyalin Island, people live on it"	"Second Inyalin Island, people live on it"
17. "Map which shows discoveries of Russian seamen on Northern part of America with nearby places on various voyages which occurred." I.F. Truskott - G.F. Müller 1773. AAN, p. IX, op.la, # 210; ATsKP VMP, #86.	-	-

Islands		
Fairway Rock	King Island	Cape Prince of Wales
(Abridged) "Empty Is. called Okivahai"	(Abridged) "Okiben Island. There live the same people as on the Big Land"	Cape Prince
Without a name	-	"Big Land"
Without a name	Without a name	Land discovered during geodesist Gvozdev's voyage From Kamchatka
"Child without a home Okivahai"	Without a name	"Land called Kygmyn"
-	-	"Coast discovered by geodesist Gvozdev in 1730"

Source	Ratmonov	Diomede Kruzenshtren
18. Copy of N. Daurkin's map (1765) with reviewed explanation, 1774. TsGADA, f.199, portf. 539, p.1, d. 1a, l. 20.	"Imaglin Island"	"Inyalin Island"
19. "Chart of Norton Sound and Bering Strait..." Cook and Clerke, 1778-1779, (Published in English 1782, 1784, 1785).	-	-
20. "Map belonging to sotnik Ivan Kobelev's voyages" 1779 *Meshyatolov Historical and Geographical*, St. Petersburg, 1784.	"Imahlin"	"Igellin"
Handwritten version of the map of I. Kobelev's voyage in 1779. Federov, 1971, ill. 1.	"Imaglin Island"	"Igellin Island"
21. Map of 1781 from the "Collection of various maps indicating route to Arctic Ocean and America" AVPR f. 339, op 88 No. 930/10; TsGVEA, f. VUA, No.23741, i.1.	Without a name	Without a name

Islands		
Fairway Rock	King Island	Cape Prince of Wales
"Empty Okivahai Island"	"Okiban Island"	"...Big Land, part of America, and called in Chukchi Kygmyn"
-	"King"	"Cape Prince of Wales"
Usien"	-	"American Nos was discovered by geodesist Gvozdev in 1732 and named Gigmalskaya land"
"Ukzen"	-	"American Nos which was known in 1730 as Kygmalskaya by Geodesist Gvozdev"
Without a name	-	"American Nos, on which Gvozdev has been in 1732"

Source	Ratmonov	Diomede Kruzenshtren
22. "Eastern portion of Irkutsk Province with nearby island and Western coast of America" Vilbrekht, 1787. All names are repeated without changes in Russian and French publications of the "maps showing the discoveries by Russian seamen in the Pacific Ocean and English Captain Cook" Vilbrekht, 1787. TsGVIA, f. VUA, No. 23780, 23758.	Without a name	Without a name
23. "Merkatorskaya map, indicating Arctic Ocean, Bering Strait and part of the EasternOcean with the coastof Chukota and North America" Sarichev, 1791. Sarichev 1802 Atlas, i. 52.		"Gvozdev Islands"
24. Merkatorskaya map of Bering Strait" Kotzebue, 1816 Kotzebue, 1821	Ratmonov	Gvozdev Kruzenshtern
25. Merkatorskaya map, indicating Arctic Ocean, Bering Strait and part of the Eastern Ocean with coast of Chukota Land and North America" M.N. Vasilev and F.P.Wrangell, 1821. AVPR, f. 339, op. 888.	Without a name	Gvozdev Without a name

Islands Fairway Rock	King Island	Cape Prince of Wales
Without a name	King Island	"C. Prince of Wales. This nos was already known in 1732 by geodesist Gvozdev by name of Kygmalskaya Land"
"Kivahoi"	"Okiben"	"Cape of Kigmile Village"
Islands Without a name	"King"	"Cape Prince of Wales"
Islands Without a name	-	"Cape Prince of Wales"

Source	Ratmonov	Diomede Kruzenshtren
26. "Map of Russian possessions in N. America" V.N. Berkh, 1821, ATsKP, VMP, No. 188.	Without a name	Without a name
27. "General map of Bering Sea" and "Merkatorskaya map of Western coast of Bering Strait, made from description of Fleet-Captain F. Lütke from the sloop *Sinyavine*, 1828, Lütke, 1835, Atlas, I. 1,2.	"Nunarbook"	Gvozdev "Ignaluk"
28. "Meratorskaya general map of part of Russian possessions in America" L.A. Zagoskin, 1844 (Published in 1848) Zagoskin, 1956, appendix.	"Imaklit Island"	Gvozdev Islands or St. "Inaklit Island"
29. Sea Atlas, 1950-1958.	Ratmonov Island -Big Diomede -Imaklit 65°45′N.L. 168°55′W.L	"Diomede (Gvozdev's) Kruzenshtren Island -Little Diomede -Nurarbook 65°45′N.L. 169°00′W.L.

Islands Fairway Rock	King Island	Cape Prince of Wales
Without a name	Without a name	"Cape described by Gvozdev in 1732"
Islands "Ukiyok"	"King's Island"	"Cape Prince of Wales"
Diomede "Ukiyok Island"	"Ukivok Island"	"Cape of Nihta Village or Cape Prince of Wales"
Islands "Fairway Rock" 65°45'N.L. 168°45'W.L.	"King Island" 64°59'N.L. 168°01'W.L.	"Cape Prince of Wales" 65°35'N.L. 168°05'W.L.

Index of Names

Alaida Is. 97
Alazeiski Mt. 38
Alazeiski River 38
Aleutian Is. v, 120, 129
Alphimov, F.D. 11
Amur River 33,97,98,132
Anadyr 15,16,18,28-34,38-9,41-2, 47,54-5,66,68,70,73,80,94, 120-1,124,133,145
Anadyr Detachment 31,38,40, 117
Anadyr Gulf 120
Anadyr River 39,42-3,47,93-4, 123,150
Andreev, A.I. 92,137
Apraksin, F.M. 6,8
Apushkin, F. 8,9
Aritlander, U. 100-1,135
Arkhangel Mikhail 89,95,97-9
Arkhangelskai Province 12
Arsenev, S.A. 11
Astrakhanskai Province 20
Atlantic Ocean 125
Avacha River 44
Averstev, I. 75
Balashov, I. 11,13
Baluev, F. 11,13,14
Baskakov, D. 20
Batom River 100
Baturin, B. 12
Beliaev, A.A. 9
Beliaev, G. 29
Belousov, P. 135
Belov, M.I. 52,137
Berg, L.S. 52,54,56,127

Bering, V. v,14-5,19,31-2,39,48,54 74,82-3,85,89,92,94,102,106, 108,109,113,118-9, 123,127, 137
Bering Expedition 16,43,54,108, 117,121,128-9,137,143-4
Bering Sea v,15-6,31,120,138,158
Bering Strait 54,62,92,119-21,125- 6,128,137-8,140,142,146,148, 154,156,158
Berkh, V.N. 34,126,129,138,158
Bibl, E.B. 136
Big Land 1,33-4,36,39,43,47-9,60- 67,69,70,73,92-5,103-4,106-7, 118-21,125,128,131,133,147 149,153,155
Biltsov, P. 6
Bobovsky, I. 99,101
Bolshaia River 34,40,76,81,129
Bolsheretsk 32-3,37-8,40-3,54,70, 73,76,79,80-2,95-6,124,131-2, 135-6
Bolsheretsk 95
Boltin, I 28
Borisov, M. 81
Borodavkin, K. 4,13,20
Brill, A. 115
Browner, P. 101
Bukharin, Y. 6
Bykov, I 4
Cape Anadyr 14,47,54,70
Cape Chaplin 54
Cape Chukotka 54,60,65-6,94, 125
Cape Dezhnev 54,56,128

Cape Kingegan 63
Cape Lütke 54
Cape Prince of Wales 63,123, 126,131,147,149,151,153, 155,157,159
Cape Spencer 60,144
Catherine the First 14
Chaplin, P. 32
Chekavka River 96
Cherkaski, A.M. 15
Cheruiski 27
Chetverikov, M. 6
Chichagov, P.I. 4,20
Chichagov, S. 9
Chicherin, I. 4,20
Chikin, N. 115
Chirikov, A.J. 110-1,113,119, 122-4,129,139
Chukchi 34,54-6,64,66,92,94-5, 118,126,138,146-7,149-150, 152,155
Chukotka 15-6,34,39,59,65,69, 92,107,121-3,138,135,138-9 141,148-9
Chukotka Peninsula 16,39,40, 54,59,62,92-3,113,118-121, 123-5
Chukotka Expedition 40
Columbus vi
Commander Is. 120
Cook, J. 125-6,137,143-4,154, 156
Cossack 2,14,19,27-36,38,41,51, 61,65-7,76,115,126,138,146, 161
Czarina Sophia 8
Danilov V.A. 72

Dasajev, Y. 102-3
Daurkin, N. 112,152,154
Davydov, Colonel 13
Delisle, J.N. 118,120-1
Devier, A. 39,66,90-5,100,105,134
Divin, V.A. 52-3,107,139
Djagilev, I. 101
Dobrynski, I. 81-2
Dufua, A. 127
Dyakov, C. 20
East (Pacific) Sea v,vii,28-9,31,44, 62,117-8,120,123,125,128,139 140-4,156
Eastern Gabriel 32-3,36-8,40,53
Eastern Siberia v,44,114
Efimov, A.V. 52,54,64-5,71,92, 104-5,107,139,140,146,150
Egache River 34
Elizabeth II 99,125
Eniseisk 28,111-3,132
Eniseisk Provicne 27,111
Eselberg, A. 102
Ezhevski, F.K. 13
Fedorov, I vii,17,19,27-31,33,35-6, 38-41,47-9,51-3,59,62-4,66,68, 69,70,93,99,103,105-7,113, 119,133,135
Fedorovich, P. 75
Fenev, A. 86
First Kamchatka Expedition v,14, 31,39,54,119,128
Fortuna 31-2,42,79,89
Gardebol, S.
Geinzius, I.G. 6
Gens, Y. vii,17,19,27-8,30-1,33-4, 35-42,45,47-9,51-2,70,73,74-9, 82-5,89,119,133

Gens, Maria 83
Gerasimov, A. 4,12
Gerasimov, I. 44
Gerasimov, T. 101
Gigmalskaya Land 126,155
Ginter, E. 102,111,113
Gizhiga River 34,40,47
Goldenberg, L.A. 45,138-9
Golder, F. 51,64,109,117,144
Golovnin, V.M. 126,138
Great Northern War 10
Great, Peter the, see Peter
Greis, R. 3
Grekov, V.I., 61,105,139
Grigorjev, P., 111-2
Gvin, S., 3
Gvozdev Islands, 128,138,156, 158
Gvozdev Cape, 129
Homann, I. 15,18,24
Igellin Island 61,154
Ignatev, S. 12
Ignatev, M. 11
Ilimski 27-8
Imaglin Island 61,154
Imaklik Mt. 71
Imiev, A. 78
Inatlik, Mt. 71
Indigirka River 38
Inia River 91
Irkutsk 14,27-8,33,39,70,81-4, 92-5,102-3,105-7,108,111, 115,132,135
Irkutsk Province v,69,83-4,89-90, 109,115,146,152,156
Isakovs, I., 10
Isupov, A. 6

Ivanovna, Anna 89,134
Ivashkin, S. 101,111-2
Izamailov, L. 11
Japan 91,95-6,98,140,152
Javile 128
Kamchatka v,15-6,24,28,31-2, 34-6,38,40-5,47,53-4,60,64,67-70,73-4,76,78-82,85,103,112, 119,121,124,131,133,137-40 142,148,150,153
Kamchata Cape See Lopatka Cape
Kamchatka Detachment vi,47, 73-4, 79,80,86,131
Kamchatka River 32,40,42-4, 53-4,68,76,78,93,107-8,119, 120,125,131,135,152
Kamchatka Exped. of 1754 124
Karaginski Island 125
Kashintsov S., 13,20
Katashetsov I., 70
Kaviaiak Bay 63-4
Kazanskai Province 12
Kazantsev, V.I. 62,69,73-7
Khalerchinska Tundra 38
Khanykov, I.F. 10
Kharchin, F. 43
Kharuzena River 44
Khitrov, S. 82
Khmylev O., 33
Khrushov I. 4,12
Kievskai Province 17
King (Ukivok) Island 64-6,126
Kirenski 27
Kirilov, I.K. 13,15,18,24,122,154
Kobelev 61-2,126,129,154
Kolyma 15,31,39

Kolyma River 15,31,39
Kolymski 29
Koryak 32-3,35,60,70
Korostelev D. 99,100,102,135
Kostiurin, P. 6
Kostochkin, N. 11
Kotzebue, O.E. 127,129,140,156
Kozin, M. 95
Kozlov, F. 32
Kozyrevsk, I. 28-9,45
Kraft, G.V. 6
Krapivin, I. 12
Krasilnikov, A.D. 9,11
Krenitsin-Levashov Expedition 124
Kriskov, I. 43
Krivolutski 27
Krivov, A. 6
Krondstadt 84
Krotkov, A. 13
Krug, P. 72
Krupyshev, T. 35,80
Kruzenshtern Is. 65,123,146, 148,150,152,154,156,158
Kruzenshtern J.F. 127
Kuchin, F.F. 11-2
Kuchin, S. 12
Kudinski 115
Kudrin, G. 11
Kuhtui River 120
Kurile Islands v,14,16,28,32-3, 95,97,118,120,124,129,132, 134-6
Kuzmin, A. 89,102,111
Kuznetsov, S. 112
Kykchig River 40
Kykhmyltsy, 66

Laborovski, A. 83
Ladozheski Canal 11
Lama 41
Lamuts 33
Lamskoe Sea, see Okhotsk Sea
Lang, L. 39,92-5
Laperouse Strait 98
Lavrov, F.D. 13
Lebedev, I. 20,33
Lena River 28-9,31,38,45
Leningrad vi,1,3
Leushinski, V. 4
Lindenau, I.A. 39,148
Lion 32-3,43
Lodyshenski, T. 4
Lomonosov, M.V. 9,125
Lopatka Cape 43,54,119-21
Lovitsa, G.M. 128,148
Lupandin, P. 22
Lütke, F. 127,140,158
Luzhin, F.F. 20
Lvor, I. 122
Lykov, S. 11
Ma River 28
Magellan vii
Magnitski, L.P. 3-6,9
Makarov, G. 20
Makarov, V. 31
Malchikan 91,100,134
Malchikan River 91,132,135
Malyshev, A. 67
Menshikov, A.D. 74
Merlin, V.F. 73-4,81-2
Michurinski Plan 13
Molchanov, F.U. 4,11
Mordvinov, D.A. 10
Moscow vi,13-4,20,23,27,84,131

Moscow Province 12,20
Moscow Road 115
Moscow Mathematical-
Navigational School 1-7,11,20,
131
Moshkov, K. 51,53-4,59,68
Müller, G.F. 1,27,34-5,93,117,
120,123,127,129-30,141,144,
150,152
Myatlov, V.A. 124
Myshetski, A. 11
Nadezhda 89,95-9,101,129,132,
134-5
Navy Special Education
Corps 4
Nazarev, A. 75
Nekhoroshikh, G. 67
Nerchinsk Expedition 61,114-5
Nizhnegorodskai Province 12
Nizhnekamchatsk 39,42-3,47-8,
68,73,75-6,79-82
Nizhneostrozhnye Dept. 76
Nizovaia Expedition 20
Nome 64
Nome Cape 64
North America v,61-2,64,117-8,
120,125,131-5,156
North (Arctic) Sea 20,117,123,
139,141-2,144,148,154,156
North Star 18
Norton Sound 64,154
Novgorod 2,12-3,131,133
Nuukan 60
Okhota River 36,89,91-2,98,100,
116,132,134
Okhotsk v,29-33,35-9,44-5,51,

Okhotsk (cont.) 65-6,69-70,73-7,
78-83,85,89,90-95,97-103,105-
8,111-3,117,119-22,124-5,129,
131-5,148
Okhotsk Road 39
Okhotsk (Lamskoe) Sea v,14-6,
31-3,38,78,91,95,118
Okhotsk Tungus 33
Okulov, G. 20
Olutor River 32
Olutorsk 78,80
Orlikov, S. 12
Ostafiev, I. 33-4
Outsin, M. 115
Ozernaia River 44
Pacific Ocean, See East Sea
Paliakov, L. 42
Pallas, P.S. 126,141,144
Paramushir Island 97
Paranchin, F. 42,67
Paren River 34
Partovskimi Detachment 76
Pavlutski, D.I. vii,2,24,27-31,35-
41,47-8,51-2,68-70,73-4,80,82,
85-6,95,103,117-8
Pavlutski Expedition 40,42,131
Penzhin Bay 32
Penzhin River 32,34
Penzhin Sea 111,113,122,124,135,
146
Perenski Koryakski 40
Perevalov Expedition 125
Perevalov, T. 40,122,141,148,150,
152
Permiakov, E 49,61,67
Peter the Great v,1,3,8-10,14,73

Petropavlovsk 124,138
Petrovna, Elizabeth 103,115,135
Petrov, L. 28,30,33,35-7,41,49,59
   60,75-8,80,82,84-5,105,118
Petrov, M. 99,101
Petukov, I., 115
Pharvarson, A. 3-6,10-1,13
Philisov, Y. 12
Plenisner, F. 65,124,152
Plesheev, A. 83
Pochep 13
Pokhvisnev, D. 11
Pokrovski, A.A. 14,51,143
Poleshaev, K. 41
Poliakov, L. 49,67
Polonski, A. 51,93,104,142
Polujanovsky, M. 99,135
Portniagin, I. 49
Preobrazhenskaya Church 99
Preobrazhenski Department 7,8
Preobrazhenski Regiment 8,9
Preobrazhenski Village 8
Pustyshkin, I. 11
Radishev, Y. 11
Ratmanov Is.54,60,62,65,123,
   127
Ratmanov, M.J. 127
Rebrov, I. 67
Revelation 2:17
Roditchev, E. 99,102,106-8,113,
   119,122
Rogachev 89
Romodanovski, F.U. 7,8,10
Rtishev, V.R. 95-102,111,113,
   122,124,129,134-6
Safonov, E.F. 11
Sakhalin Is. v,97,129,135,142
Saltanov, S. 6
Sarichev, G.A. 127,129,142,156
Second Kamchatka Expedition
   v,1,82-3,85,89,90,95,98-9,102
   105-6,113-4,117,122-4,129,135,
   137,140
Selenginsk 115
Selivanov, S. 35,37,45
Semenovski Regiment 7,9,10
Semyonov-Tyan-Shanski, P.P.
   127
Senate 2,13,17,19,20,21-4,27-8,
   35,39,74,90,93-4,102,108,113-
   6,118,132,135
Serebriakov, S. 83
Seward Peninsula 63,119,128
Sgibnev, A.S. 19,27,30,45,142
Shaganov, A. 99,135
Shalaurova Expedition 125
Shamlev, T. 93
Shananjkin, P. 112
Shantar Islands v,33,95-6,118,
   122,129,134
Sharypov, M. 67
Shchadrin, D. 67
Shekhonski, I. 4,12
Shelting, A.E. 95-102,111,113,
   122,129,132,134-5
Shergin, A. 32
Shestakov, A.F. 2,14-7,19,24,27-
   37,39,42,51,85,118,122,131,
   133,138
Shestakov, I. 32,36,39,41,85,119
Shestakov, V. 32,36,119
Shestakov-Pavlutski Expedition
   vi,16,21,27-8,30-1,35,38,70,
   79,118-9,143,142

Shetilov, V.D. 20
Shipitsyn, V. 29
Shishkov, A. 6
Skurikhin, I.F. 33,38,51,61-2,65,
  67-8,77,82,92-5,103-4,117
Shtinnikov, A. 81
Shubinshy, A. 101
Shumshu Island 97
Siberia v,1,2,11,15,19-21,23-4,
  27-8,30,39-40,43-4,61,70,73,
  74,78,83-5,89-90,93-4,102-3
  105,111-2,114-8,124-5,129
  134,137,139-40,142-3,146
Siberian Dragoon Regiment 27
Sipiagin, A.A. 13
Skobeltsyn, P.N. 10-1,20
Skorniakov-Pisarev, G.G. 12,69,
  73-4,78-80,82-3,89-90,105,
  133
Smetanin,L. 42-3,66-7,81,83
Smolenskai Province 12
Soimonov, F.I. 4,61,115,124,139,
  150
Sokolov, A.P.52-3,71,92,104,143
Soldatov, O. 68,80
Somov, V. 22
Sophino Village 13
Spanberg, M.P. 1,54,63,65,74,83
  85,89,91,93,95-108,111-2,
  118-20,122-5,132,135-6,148
Speshnev, I. 27-8,30-1,33-9,41-5,
  47,49,74-83,92,113
St. Diomede Is. 54,58,62-3,65,
  120,123,126-8,130,138,146
  148,150,152,154,156,158,159
St. Joann 95,97-9
St. Gabriel v,26,32-4,37-8,40-3,
St. Gabriel (cont.) 45,47-9,51,53-
  4,59-61,63-4,66,68-9,80-3,85,
  89,93,108,117,119-20,125,
  127-8,131-3,135
St. Lawrence Island 120
St. Paul 129
St. Peter 119,129
St. Petersburg 14,17,19,23,27,33,
  39,44,84,92,106-8,113-4,118,
  122,131,154
St. Petersbury Naval Academy
  1-2,5-6,12,20,118,131
St. Petersburg Province 12
Stromilin, M. 14
Stupin, O. 75
Sukhareva Bushnia 3
Sumarokov, N. 9,13
Surovtsov, V. 20
Svistunov, I.S. 10,20,98
Tai-Khaiakhtakh Mt Ridge 38
Tambov 14
Tarabykin 35-6
Tashiverski 29
Tatarinov, M. 62
Tauisk 33-6,80
Teglev, F.I. 11,20
Terpeniya Cape 97
Terpaniya Gulf 97
Thadoviski 74
Thikhmanov, A. 9
Tibalov, F. 51
Tigil River 44
Tobolsk 27-30,38-9,66,69,70,83-4,
  89,92
Tobolsk Province 2,86
Tolubeev, A. 10,20
Tomsk 111-5,132

Trapezinikov, F. 125
Treska, N. 33,36,42
Truskott, I.F. 123,129,152
Tuev, A. 83
Tuev, I. 38
Tugur River 97
Tui 36
Turnaev, M. 29
Tylka River 34
Ud 73
Uda River 91-2,96-8,122,129,132 143
Udsk 29,92,96
Uka River 31,33,96
Ukivok Is., see King Is.
Uliia River 91
Urak River 33
Urenev, S. 11
Uritski 115
Ushakov, V. 6
Ust-Kut 28
Utka River 40
Vakhla River 34
Vasco de Gama vi
Vedrin, Y. 11
Vereshagin, I. 102
Verkhnekamchatsk 43,75
Verkni 77,79,81
Verknoiianski Mt. Ridge 38
Vertlugov, P. 125,152
Vize, V.Y. 62
Vinsgeim, H.N. 6
Volkov, M.Y. 2,12-3,131

Voronezhskai Province 14,20
Voronov, F. 112
Vsinov, V. 115
Walton, W. 91
Waxell, S. 93,113-4,122
Wilbreht, A. 126,129
Wolf 115
Yakovlev, V. 4,9,12
Yakuts 33
Yakutsk 14-6,19,27-36,38,40-1, 68,70,73-4,79,83-5,91,97,99, 103,131,132,143,150,152
Yakutsk Regiment 99
Yan River 38
Yatskov, V. 38
Yudoma River 28
Yudoma Cross 29,31,33,73,97
Yudom 103
Yushin, K. 99-100,102,106-8,113, 119,121-3,132,135
Zagoskin, L.A. 127,140,158
Zaharov, I. 20
Zalevin, I. 67
Znamenski, S.V. 140
Zhigansk 74
Zybin, A. 93-4,99-100,102-3,105- 108
Zinovev, K. 11
Zubov, F.A. 10
Zurov, V.M. 11
Zvezda, P. 45
Zyrain, V. 67